REFRAMING THE ENVIRONMENT

This volume unravels the underlying power relations that are masked in the present discourse of ecological sustainability and conflicts over natural resources. Current discussions on environment emphasise the use and abuse of the environment in various ways. This book looks at the inter-linkages of discourse, resources, risk and resistance in the contemporary neoliberal world. While exploring the experiences of neoliberalisation of nature in India, it brings out the intersections of conservation and management, science and gender, community politics and governance policies.

The volume highlights the cultural politics of resistance from multiple sites and regions in India in the recent context (be it land, water, forest, flora or fauna or urban commons). It discusses the ways in which environmental issues have come up and been appropriated, while examining the role of the State and actors such as corporates, traders, consultants, ecotourism companies, green activists and consumers, and consequences of 'green' appropriation and the 'growth' story. The major themes of the volume are the interrelations of nature, culture and power; neoliberal governance and the environment; access to and use and management of land, natural resources and environment; community politics and livelihoods; marginalised groups and local communities; marketisation and the environment; and new forms of re-appropriation and resistance.

This book will be of great interest to students and researchers in sociology, environmental studies, environmental history, environmental anthropology, political ecology, political science, geography, law and human rights, economics and development studies as well as to environmental activists, policy makers and those in media and journalism.

Manisha Rao is Assistant Professor at the Department of Sociology, University of Mumbai, India and previously taught at the Department of Sociology at SNDT Women's University, Mumbai, India.

REFRAMING THE ENVIRONMENT

Resources, Risk and Resistance in Neoliberal India

Edited by Manisha Rao

Routledge
Taylor & Francis Group
LONDON AND NEW YORK

First published 2021
by Routledge
2 Park Square, Milton Park, Abingdon, Oxon OX14 4RN

and by Routledge
52 Vanderbilt Avenue, New York, NY 10017

Routledge is an imprint of the Taylor & Francis Group, an informa business

British Library Cataloguing-in-Publication Data
A catalogue record for this book is available from the British Library

Library of Congress Cataloging-in-Publication Data
A catalog record for this book has been requested

ISBN: 978-1-138-23766-7 (hbk)
ISBN: 978-0-367-55318-0 (pbk)
ISBN: 978-0-367-55316-6 (ebk)

Typeset in Sabon
by Apex CoVantage, LLC

To my supervisor – **Professor D. N. Dhanagare** – from whom I learnt the importance of dedicated hard work in academic reading and writing

To my teacher, mentor and dear friend – **Sharmila Rege** – for shared journeys of exploration in academics, activism and life

CONTENTS

CONTRIBUTORS

Gita Chadha is Assistant Professor at the Department of Sociology, University of Mumbai, India. Her academic interests include sociological theory, feminist science studies, sociology of knowledge, intersectional feminist epistemologies and visual cultures. She has designed and taught the first feminist science studies course in India at Tata Institute of Social Sciences, Mumbai. She has developed frameworks for feminist archiving at the Research Centre for Women's Studies, SNDT Women's University and has designed interdisciplinary pedagogic initiatives for science and social science teaching. Her latest book publication is *Re-imagining Sociology in India: Feminist Perspectives* (edited with M. T. Joseph, 2018).

Hemantkumar A. Chouhan is Assistant Professor at the Department of Sociology, Savitribai Phule Pune University, India. His research and teaching interests pertain to sociology of environment, sociology of development and urban sociology, sustainability and livelihood issues of fisher communities.

Arun de Souza was Lecturer at St. Xavier's College, India where he taught sociology and anthropology. His disciplinary interests are environmental and development anthropology. His PhD work on watershed development has been published as *Water and Development: Forging Green Communities for Watersheds* (2010).

Ritambhara Hebbar is Professor and Dean at the School of Development Studies, Mumbai, India. She specialises in tribal studies, specifically tribal culture, movements for self-rule and governance. Her publications include her book *Ecology, Equity and Freedom: Engagement with Self-Rule in Jharkhand* (2011) and the co-authored book *Towards a New Sociology in India* (2016). Her recent research is on security guards in Mumbai. She is also engaged in research on tribes in South India. She teaches on related themes such as understanding Indian society, tribes in the contemporary world, tribes in India, qualitative research approaches and methods, methodological issues in development research, social exclusion in India, environment and society, anthropology and development.

Seema Kulkarni is Senior Fellow and one of the founding members of Society for Promoting Participative Ecosystem Management, Pune (SOPPECOM), India. She co-ordinates the gender and rural livelihoods activities within the organisation. She has co-ordinated various studies and programmes around decentralisation, gender and land, water and sanitation. She has published several articles and book chapters around issues of gender, water, sanitation and rural livelihoods. She has been associated with Stree Mukti Sangharsh Chalwal, the movement for the rights of single women in Western Maharashtra, India and is involved in the coalition of women's groups in Maharashtra, Stree Mukti Andolan Sampark Samiti. She is currently the National Facilitation Team Member of the Mahila Kisan Adhikar Manch (MAKAAM) forum for women farmers' rights, and anchors the network at the Maharashtra level. The national secretariat for the network is also currently co-housed in SOPPECOM.

John Kurien is Visiting Professor at Azim Premji University, Bengaluru, India and retired as Professor, Centre for Development Studies, Trivandrum. He is also a practitioner and has worked closely with small-scale fisherfolk in Kerala, Cambodia and Indonesia.

B. Manjunatha is Research Associate at Azim Premji University, Bengaluru, India. He has considerable experience with protection of urban commons in Bangalore, and has a number of publications in this area.

Harini Nagendra is Professor at Azim Premji University, Bengaluru, India, where she leads the Centre for Urban Ecological Sustainability. She is also the author of *Nature in the City: Bengaluru in the Past, Present, and Future* (2016). She is an ecologist who uses satellite remote sensing coupled with field studies of biodiversity, archival research, institutional analysis and community interviews to examine the factors shaping the social-ecological sustainability of forests and cities in the South Asian context.

D. Parthasarathy is Professor at the Department of Humanities and Social Sciences, Indian Institute of Technology, Bombay, India. He is Convener, Interdisciplinary Program in Climate Studies, IIT Bombay and Associate Faculty, IDP in Climate Studies, Centre for Policy Studies, and Centre for Urban Science and Engineering. His research interests include urban studies, development studies, law and governance, legal pluralism, vulnerability and adaptation to climate change, gender and development and disaster studies. He has published extensively in national and international journals. He has co-edited with Tim Bunnell and Eric Thompson *Cleavage, Connection and Conflict: Rural, Urban and Contemporary* (2013).

Manisha Rao is Assistant Professor at the Department of Sociology, University of Mumbai, India. Her academic interests include sociology of environment, gender studies, sociology of development and globalisation and the sociology of social movements. She has published articles on similar themes in national and international journals. She was a guest editor for the Special Issue of the *Sociological Bulletin*, 'Negotiating Neoliberal Environments in India' (Vol. 67, Issue 3, December 2018). Some of her published papers include 'Interrogating the Sociology of Environment in Western India: A Gendered Understanding', in *Genders, Feminisms and Sociologies: Towards a State of Altered-ness* (2018). Her article 'Gender and the Urban Commons in India' is published in the *International Quarterly for Asian Studies* (2020).

Geetanjoy Sahu is Associate Professor at the Centre for Science, Technology and Society, School of Habitat Studies, Tata Institute of Social Sciences (TISS), Mumbai, India. His research and teaching interests include environmental jurisprudence, environmental regulation and policy, forest rights and governance, the political economy of public policy and institutions and environmental movements. His publications include the book *Environmental Jurisprudence and the Supreme Court: Litigation, Interpretation and Implementation* (2014).

Arupjyoti Saikia is Professor of History at the Indian Institute of Technology-Guwahati, India. His research and teaching interests include economic, environmental and political history of modern Assam. His latest publications include *The Unquiet River: A Biography of the Brahmaputra* (2019) and 'Geographical Exploration and Historical Investigation: John Peter Wade in Assam' in *Landscape, Culture, and Belonging: Writing the History of Northeast India* (2019).

Chandan Kumar Sharma is Professor, Department of Sociology at Tezpur University, Assam, India. He was educated at Cotton College, Guwahati and Delhi School of Economics, Delhi. His research interests include issues concerning development, environment, migration, identity politics, social movements, agrarian change and urbanisation with special reference to northeastern India. Besides his academic publications, he has been writing on these issues for various newspapers and journals in English and Assamese. His columns also appear in Hindi newspapers. He has been a visiting fellow to Delhi School of Economics, Jawaharlal Nehru University and a Charles Wallace Visiting Fellow to Queen's University, Belfast.

Hita Unnikrishnan is a Newton International Fellow at the University of Sheffield, UK. She is also visiting faculty at the Centre for Urban Ecological Sustainability, Azim Premji University, Bengaluru, India. Her research has explored the changing nature of ecological and social vulnerabilities of urban lake social-ecological system in Bengaluru. She is also interested in historical narratives of change around the governance and appropriation of urban commons. Her work combines ecological field studies with archival research and community interviews to examine changes in urban commons in the global south.

Sudha Vasan is Associate Professor at the Department of Sociology, University of Delhi, India. Her research and teaching focus on the intersections between society and ecology. Her publications range from ethnographies of state and community forestry institutions, customary and modern forest laws, intellectual history of forestry knowledge production in India, wildlife tourism, youth culture and identity, social capital and reproduction of gender and caste inequalities to theoretical engagement with the ecological crisis and the logic of capital. The social and ecological politics of the Himalayan region remain an abiding interest and field for her writing, including a book titled *Living with Diversity: Forestry Institutions in the Western Himalaya* (2007). She obtained her PhD from the Yale School of Forestry and Environmental Studies and has held post-doctoral fellowships in the London School of Economics, Australian National University and the Indian Institute of Advanced Study, Shimla. Her research has been published among others in journals such as the *Journal of Development Studies*, *Economic and Political Weekly*, *Environmental Research Letters*, *Sociological Bulletin* and *Conservation and Society*.

FOREWORD

To claim that we live in a time of epochal changes may sound trite, but the transformations are so all-pervasive that we take notice of them. As much and perhaps even more than demographic expansion, the pace and scale of material changes due to the growth of production is central to vast changes on a planetary scale. Historians of environmental change differ on when the human interface with earth's living systems underwent a fundamental shift. Some place this in the period 1500 when global contact and expansion driven by the West opened up a new era, once labelled the Vasco Da Gama epoch. Others see the end of the Second World War as a time followed by significant changes in technology which in a short time created not only more wealth but also more waste than ever before in human history. It is no coincidence that India or the sliver of the lands and waters that made India was and is a central arena and theatre. It is therefore appropriate we come together to ask how best to study these changes in our own times. India may be the focus but the issues pertaining to socio-ecological, cultural and political contests are of much wider significance.

This is by no means the first collection on ecology and society in flux in India. But earlier works have taken up specific issues such as forests or water or looked at contests over commons or the nature of the law or of policy. Over a quarter century ago the pioneering duo of the ecologist Madhav Gadgil and social historian Ramachandra Guha posed the issues at a time economic reform was at incipient stage. Today, India has seen a growth rate of 6.5 per cent for nearly four decades (since 1980) and a deeper integration with global systems of exchange and movement of capital, goods, services and people. The larger role of governments of whatever hue (especially since the end of the 1980s) has become more of a facilitator, not a source or driver of fresh investment.

That this has come at a time of greater global and national concern of the close, intimate connections of natural systems and human well-being is also striking. That the glaciers of the Himalayas are water towers of South Asia is not new but the concerns about global climate change are more acute now than before. The death of species long excited attention but the extirpation of entire water bodies whether lakes or rivers makes it an issue of survival not mere aesthetics. Beyond specific ecotypes such as forest, mountain or riverfront, there is now evidence of far reaching and interconnected processes often under strain due to the levels and kinds of human intervention. Repair, renewal and resuscitation are possible and do occur but are often unable to keep pace.

Interestingly, one of the key arguments for valuing nature is to preserve it, but in an age of capital, it is critical to ask if this is a response that is workable even in its own terms. Or is it a way to deeper tragedy not only for those reliant on nature for living but even for those who gain in the short term? Conversely, will short-term imperatives of accumulation undermine legislative, executive and community-led attempts at regulation? Is the crisis of ecology to be driver for reform or for more radical larger ideas of social transformation? Is it to be one or the other or perhaps *both*?

The rise of India – and for that matter China – as global economic players has deep, far-reaching consequences for hydrology and soil systems, the life forms and living systems of biomass production that have long sustained humans in these vast societies. In the Indian case, these issues of ecology and equity acquire sharper form both due to the reality of democratic politics and the persistence of undemocratic and unequal modes of power. Unlike in China, the spaces for the former are real, if contested, keeping alive hopes for strong but non-violent modes of resistance and reform. While the emphasis here is on neoliberal reforms and their underside in socio-ecological terms, there are also deeper continuities. There are threads that link us to the imperial era (as with laws on forests or fisheries or with the Denotified Tribes and itinerant peoples) and with princely India (as with former hunting reserves now national parks). There are other continuities with the early years of independence (that saw heavy industry and large projects take centre stage) or with land appropriation for national projects (under a law enacted in 1894 and replaced as recently as 2013). This kind of continuity is by no means unique to India. What makes it relevant is that even as older struggles have not died away, new foci of concern have also arisen.

It is vital here to set the quest for democratic alternatives apart from and in distinction with more statist and market-driven approaches. The use of nature by some to exercise power over society is by no means a hallmark of the modern era. After all, royal control over the forest was vital for war material (elephants and timber) as well as for hunts (a surrogate for war-making). Long before fossil fuels, the inequitable control of human and animal labour power and energy was critical not only to state formation but to the appropriation of the wealth of nature by dominant groups and classes. Closer to our times, no less that Rudyard Kipling celebrated the foresters, engineers and civil and military officials who imposed order on subject lands and peoples and kept the flow of wealth in timber, cotton and grain flowing over bridges and along canals and rail networks. If one of his creations, Kim was an undercover agent; the other, Mowgli enlisted as a forest guard.

To cut a long story short, independence opened up historic opportunities for change, but one marked feature was that both extraction of wealth and the protection of nature was often at the cost of those who were more reliant on nature. Thus, conservation by state fiat of rare species (the rhino or the elephant) or of entire ecological systems (the Doon Valley or the Yamuna river front) or large classes of landscape (the state forests as per the Godavarman judgment of 1996) perpetuates exclusion to protect nature. It is – as with the imperial or royal reserve – certainly a form of protection, but it is equally a form of appropriation. What the environment means can differ greatly depending on who you are, what you do for a living and where you live. It is for this precise reason that the issues of the environment, of living spaces and resources are so deeply contested. Freedom and liberty go well together, but what is its record on equality like more so in the age of reform and globalisation? Wealth is created as well as waste generated in ways that have very different meaning for different sets of players.

This raises a key question: how can a democracy sustain a peace with nature by just means? Any serious response requires a wider debate on the lineages and contours of the present. This anthology and the larger conference it grew out of draws on historical debates and ecological studies, but it gives them a sharp focus around the contemporary condition. Where history asks how the past shapes the present and the ecologist probes the many ways in which the structures and functions of nature work, sociologists have in many ways been well ahead in India and other parts of Asia in their studies of the interlinkages of social formations, political relationships and economic processes with environmental change in the widest sense. The transformations of land, water and atmosphere unmistakable in Indian metropole or small town do pose a set of new challenges. But as works here on Brihan Mumbai and Bengaluru, the financial and software powerhouses today show, the issues of living by fishing or drawing on lakes for livelihood remain critical and contested. Just so, the integration of the oceans into global networks of exchange is not only transforming marine systems faster than any time in half a millennium, but also posing new challenges to the fishers' movements for dignity and livelihood all along the coast. It is not nature alone that is contested; it is the ways in which the social fabric is being remade in ways that squeeze out those at the margins, even as they make the repair and regeneration of nature increasingly uncertain if not impossible. Though much abides, much is taken.

How these stories will unfold is unclear but a parallel may help. At any point of time in history, changes are complex and multi-faceted. Not included in this collection is the talk by veterans Gail Omvedt and Bharat Patankar who drew on the work of the radical social reformer Jyotiba Phule, who saw the late 19th-century imperial encounter as multi-faceted. Education, army and public jobs enlistment made upward social mobility a reality for the majority who lived by labour and whose occupations were tagged as unclean. But the struggle for dignity and rights had to also take account of the expanding powers of merchant capital, the resistance of entrenched educated elites and the new ways in which government appropriated forests and pastures. The issues of universal franchise or land to the tiller, the rights of women and the so-called lower social orders were central to the 20th century. Just so, the early 21st century sees both new enclosures of not only common spaces and of rights, but also witnesses new and innovative ways of expanding precisely the means by which dignity and livelihood can be asserted, protected and made more, not less, secure. Works here such as by senior scholars of fishers or of the peoples of the Nilgiris, and the questions of coercive megafauna protection and dam building in the North East shed new light on continuing concerns. As in Phule's time so too in our own, there is no one way forward, but there are common threads of concern.

The issue of gender and its relation to the politics of science is one such thread. Whereas in earlier eras the latter was seen fundamentally as emancipatory, it is striking how even the massive movement of land reform reduced the issue to one of redistributing land to the male head of household. This larger setting may well explain how the questions of gender remained marginal to the development debate until well into the 1970s. It also helps focus at another level and stratum of society on how the larger processes of technological intervention and science-based modernisation did not more fully engage with traditions such as those of J. C. Bose. Just as with the wider penumbra of movement of social and economic reform, there is a wealth of material on the past that will help enrich our engagement with the present. Issues of patriarchy or gender divisions or labour, as much as those of privilege based on birth and occupation, are now central to the politics of the environment. A mere celebration of tradition will not open up issues of exclusion or denial of dignity or rights to the lower social orders or women. But a simple embrace of modernity may paper over such cracks and enable their reinvention and revival. Nor is it simply a case of raising purchasing power or of securing participation, vital as these are. It is a call to rethink the wider issue of what the editor of the volume presciently calls the triad of 'nature, knowledge and power', but doing so in ways that make not only class but also gender matter as much.

Dialogues are as much about posing questions and continuing the quest for answers. In a sense, the former is even more vital. As this collection shows there are many layers to society and dimensions to ecology, and even a focus on a unifying themes such as the expanding reach of state structures or capital accumulation as process can take not one but a myriad forms. More than ever, India is both the confluence and starting point of not one but many enquiries. Most central to these – and this theme runs through the chapters here – is how the democratisation of states and societies is vital to ecological renewal but at odds with secular processes integral to the iterations of wealth and power at the present juncture. The quest for peace and justice hinges on our ability to go to pose and debate the hard question. This collection makes for a renewal of that effort. The Department of Sociology, Bombay University deserves all credit for having provided a platform. The issues will live on, and this book will give fresh insight in the ongoing debate.

Mahesh Rangarajan

ACKNOWLEDGEMENTS

The work of putting together this volume is the outcome of a conference organised at the Department of Sociology at the University of Mumbai, India in 2016. The theme of the conference on the environment in a neo-liberal context in India was something I had been engaged with. It related to the UGC Special Assistance Program specialisations of the department. It was also planned as one of the conferences leading up to the centenary celebrations of the department in 2019–2020.

In this regard I would like to acknowledge the support of all my faculty colleagues, administrative staff and the students at the Department of Sociology. Some need special mention: Ramesh Kamble for helping me sharpen the theme of the conference, Gita Chadha for her inputs regarding feminist intersections and encouraging support throughout the planning and execution of the conference and M. T. Joseph for his balanced advice and the solid support that he always provides. A special thanks to Professor Indra Munshi for encouraging me to think out the sub- themes of the conference, Professor Kamala Ganesh for suggesting names of scholars, Professor P. G. Jogdand for always encouraging my academic endeavours, Professor P. S. Vivek for the warm cups of lemon tea and discussions regarding university protocol, B. V. Bhosale for providing all logistical support for the conference organised during his tenure as head of the department and my colleagues B. N. Kendre, Rita Malache and Sonali Wakherde for their warm support.

The conference was partially funded by the Indian Council for Social Science Research, (ICSSR) New Delhi and by funding by the University of Mumbai.

I thank all the paper presenters, speakers, chairpersons and special invitees of the conference. Professor Mahesh Rangarajan took keen interest in the theme of the conference and made valuable suggestions. I would like to especially thank him for writing the Foreword to this volume.

Although there were a number of papers presented at the conference, not all could be included in this volume. A couple of the papers were invited specially for this volume. Some of the articles formed the Special Issue of the

Sociological Bulletin on the theme 'Negotiating Neoliberal Environments in India', Volume 67, Issue 3, December 2018.

In putting this volume together and seeing it through, I acknowledge and appreciate the immense support provided by Shoma Choudhury of Routledge, Taylor & Francis Group.

I would like to specially thank Suman Vaze, teacher, mathematician, artist and dear sister, for permitting me to use one of her brilliant paintings as the cover for the South Asia edition of this volume.

I would like to acknowledge the support provided by my family in all my endeavours, especially my mother for being the inspiration that she is. Thanks to Siddharth and Kabir for making hay while I was busy! Last but not least, I would like to thank Kushal Deb for never failing to provide his critical comments on my work and taking care of the home front!

1

UNDERSTANDING NEOLIBERAL ENVIRONMENTS IN INDIA

An introduction

Manisha Rao[1]

The spread of production and exchange relations across global networks has had important social and environmental implications. Nature is tied up not only in the material sense to commodity circuits, but is also sold for itself in its pristine form. This is reflected in the proliferation of experiencing of nature in its pristine form through travel and adventure tourism, to the sale of farm homestay experiences and so on. Nature itself is subjected to the processes of commodification. In recent years in the name of sustainability, conservation or 'green' values, there has been an increasing 'commodification' (Prudham 2009) of nature. This commodification takes various forms. The production of waste and pollution, the transformation of ecosystems like the conversion of forests into plantations and diversion of agricultural land to production of biofuels are all examples of this. Besides this, there is the movement of private companies looking for new methods to circulate capital through new ways of commodification of nature such as the commercialisation of resources like water, fish, seeds and genes. This is supported by policy changes in support of privatisation and exchange in the market. This is highlighted as better ways to protect, conserve and manage in a rational manner natural resources and the environment, leading to a broad neoliberalization of nature (McCarthy and Prudham 2004; Heynen et al. 2007).

This crisis situation has been termed 'green grabbing' (Vidal 2008). The term emphasises the growing significance of appropriation of land and resources for environmental ends. This appropriation means the transfer of ownership or use rights or control over resources that were once publicly owned into private hands of the powerful. This is achieved by calling upon green credentials while appropriating land for commercial farming to produce food (in more efficient ways so as to reduce the pressure on forests) or for the production of fuels. This is done through the alienation of land or through the restructuring of rules and authority over access, use and management of resources. However, it does not always mean an alienation of communities from their claims to the land. It is done to serve the green purpose. The phenomena of green grabbing is built on earlier forms of colonial and neocolonial appropriation of resources for environmental ends. What is new is the number of players involved who are deeply embedded in the capitalist networks and work across multiple levels and scales. It involves new forms of commodification, valuation and markets for different aspects of nature.

The chapters in this volume are the outcome of a national conference titled Reframing the Environment: Resources, Risk and Resistance in Neoliberal India[2] held in Mumbai in 2016. An attempt has been made through this introduction to trace the multiple ways and means by which the neoliberalization of nature (Heynen and Robbins 2005) occurs, the processes and policy prescriptions through which it is implemented and the negotiations and/or resistance that has been offered in India.

Neoliberalization of nature

In the past few decades there has been a shift in the agendas of government policy. Earlier, land use was regulated largely for conservation (as during the colonial period, forests were reserved ostensibly for purposes of conservation as well as for the needs of the colonial rulers); now they are regulated mainly for the market. This shift has occurred largely due to the 'neoliberal turn' in governance of environment issues that have led to the privatisation and commoditisation of nature (Bakker 2005; Castree 2008a, 2008b; McCarthy and Prudham 2004; Heynen et al. 2007). There have also been a number of publications on the neoliberalisation of environments, nature and conservation that are termed as Nature TM (Trademark) Inc. (Arsel and Buscher 2012). Nature acquires new value that is associated with global scientific discourses that have led to creation of carbon as a commodity, given green value to biodiversity, and created a sustainable market for biofuels. The shifts in policy can be seen in that earlier nature was valued for what it could offer in terms of resources or conservation and sustainable use of resources. Now it is the value of repair which, along with sustainability, encourages the repair of damaged nature. It looks at the negative effects of growth and hence emphasises the value for commodities of nature such as carbon, biofuels and so on. It is argued that unsustainable practices in one place can be repaired by sustainable practices elsewhere. Here of course power relations come into play with one nature subordinated to the other (Fairhead et al. 2012: Leach et al. 2012). In order to understand what are the new ways of appropriation of nature within the neoliberal world, one needs to look at the emerging economic order, the discourse of power relations in the global and local context as well as the material aspects in terms of the on-the-ground reality of people's lives and livelihoods.

Under the rubric of neoliberalism, there has been a continued 'accumulation by dispossession' (Harvey 2003, 2005) through the 'enclosure of public assets by private interests for profit, resulting in greater social inequity' (Bakker 2005: 543). The processes that lead to the concentration of assets of value in the hands of a few who already hold capital are privatisation, financialisation, the management and manipulation of crises and state redistributions (Castree 2005). Through these processes the neoliberal state favours capitalist business interests over others leading to accumulation by dispossession (Harvey 2003, 2005). There has been a slow but increasing trend of publicly owned assets shifting into private hands. This can be seen in the shift of ownership of common lands and grazing lands as well as farm and forest lands from the hands of the state into the hands of private companies who use it for private ends. In some cases it may also happen that the owners of the assets are too poor or have low incomes that do not support social reproduction and they are dispossessed of their assets by sale in the market.

Furthermore, nature is now increasingly being sucked into the financial system. This system ensures that nature is being properly valued. According to this new conceptualisation of nature for example, forests are no longer valued for their timber or non-timber forest produce, but are valued for the range of services they can provide to the hydrological cycle to soil systems, biological diversity and carbon storage as well as to recreational services and payment for ecosystem services (McAfee 2011). In this neoliberal age these ideas, values and practices around nature are being reconfigured by science, and this is supported by financial institutions like World Bank, international conservation NGOs or the Convention on Biodiversity (McAfee 1999; Goldman 2005; Bakker 2009). Thus, the environment becomes a business proposition that provides a dependable income from the services it provides. However, these new valuations of nature are isolated or disconnected from the material, lived realities of local communities and thus produce inequalities across scales.

Along with the financialisation of nature's assets is the construction and perpetuation of a sense of crisis. Global economic crisis is tied to global environmental crisis. This in turn affects indebted nations, particularly of the global south, who can be forced to liberalise markets and privatise public assets. Nature becomes an important asset that can be bought and sold in the market, which leads to the accumulation by some and the dispossession of the dispossessed. This can be clearly seen as happening with the mining sector where foreign direct investment is sought as well as with forest resources. These processes are supported by the state that supports capital investments by national and international investors in countries of the global south. Due to lack of strong fiscal resources, the state is forced to makes available nature's assets like land, forest and mineral resources that are leased out. Nature simply becomes a commodity that can be leased out or exchanged. To a certain extent, ecotourism too packages and sells a commodified nature and urges the ecotourist to consume a particular kind of lifestyle (Brockington and Duffy 2010).

While trying to analyze the discourse of power relations that exist between the old and the new ways of appropriation of nature, one can discern a number of continuities that exist. Earlier forms of control by the state are being continued, however now it is with new actors, both local and foreign. As Castree (2011) argues, the state negotiates across multiple interests, working both from inside and outside the economy through processes of active intervention and regulation as well as through privatisation and deregulation leading to 'green grabbing'. In order to understand how this unfolds, one must look at the earlier histories of enclosures and the different forms of territory formation by states justified on environmental or economic grounds. Removal of local inhabitants or curtailing their rights over land or resources has been a result of the creation of forest reserves and parks for the greater global good by colonial rulers in India, Southeast Asia and Africa (Brockington et al. 2008). Based on these historical legacies, environmental discourses developed that framed local inhabitants as destroyers of the environment due to their farming and land use practices. In the 1970s and 1980s the emphasis was on development and participation of the community in environmental protection. However even these were built on the earlier legacies leading to creation of uncertainties and power play. This has led to the construction of environment in relation to power and control and the idea of environmentality[3] (Agrawal 2005). Through this process rural people are either portrayed as custodians of the environment or destroyers of it without looking into the existing class dynamics. In turn nature is appropriated by a new set of actors, a new set of legal and market processes are invented and new discourses of justification are developed. This includes a variety of actors, including the state agencies, national elites, NGOs and multinational companies offering conservation schemes or international conservation organisations and tourist operators offering ecotourism schemes. Even business entrepreneurs are trying to profit from the new green capitalism by developing forest carbon offset projects and biochar companies (Leach et al. 2012). The new acquisitions of land and resources use GIS (geographic information system) technology and satellite imagery to highlight marginal lands that are acquired. These are legitimised by the global green agenda whereby land and resources are appropriated as carbon sinks, for biofuel plantations or as reserves of biodiversity. These enclosures are justified in the name of the global good or for preservation of planet earth but provide profits for companies and elite groups. The local communities are either constructed as environmentally destructive and backward, ones who need to be trained to conform with and fit the modern idea of sustainable development, or are romanticised as the original custodians of the environment. It is argued that these are ways of exercising power relations by saying that one can become a 'green custodian' if one conforms to the discipline that the market enforces (Leach et al. 2012).

The material aspects of the neoliberalising of nature can be seen in terms of the on the ground reality of people's lives and livelihoods. In some cases there is forcible removal of people from the land and restrictions laid on land and resource use practices, while in others there are long-term changes in the local agrarian dynamics with no clear-cut winners or losers. Some local actors benefit from the new green deals, however, it is not a simple equation between the state and the local community, but is complicated by gender, generation and differences of wealth. There are also resistances put up by local communities to green grabbing. However, many a time this is taken over by green marketing strategies and the media, which plays around with the local ideas and priorities and convinces some of the people that their community interests are best served by being green custodians and serving neoliberal ideas, thus creating a form of environmentality (Agrawal 2005). Many times the march of green capitalism is stymied by unruly environments – changes in water, climate, soil, vegetation, disease – that do not follow the given plan. Nature has her own mind and may not be disciplined according to market logic. This would then raise the question of who bears the costs and risks of undisciplined environments. Further one needs to ask, why neoliberalise nature? And how does this neoliberalisation of nature play out?

Politics of deregulation and reregulation

The neoliberalisation of nature consists of governance of societies, privatisation of natural resources, enclosure of commons (as well as an allied exclusion of the communities who are linked to the commons) and the prevalence of a price value. Neoliberalisation is a process rather than a thing (McCarthy and Prudham 2004: 277). It is argued that neoliberalism inherits from 'classical liberalism' the legitimacy for 'enclosure, ownership, and commodification of land, forests, water courses, and other natural resources' and that beyond this, neoliberal policies are '*about* the nonhuman world' (emphasis in original; Heynen and Robbins 2005). In other words, they understand neoliberalism as a relationship between humans and nature predicated on the features mentioned previously. It is linked to the withdrawal of state-led interventionist policies and suggests that environmentalist concerns are effective bases from where to respond to and resist neoliberal projects. Nature is detached from the complex social constraints and instead placed under the auspices of the market leading to an 'accumulation by dispossession' (Harvey 2003).

Castree observes that the neoliberalisation of nature is understood in ideal-typical forms: apart from the stress on the privatisation of the non-private, marketisation of the non-market and deregulation of the heretofore-regulated, there are 'reregulation' (deploying state policy for the benefit of privatisation and marketisation), 'market proxies in the residual public sector' (the privatisation of the ethos of public sector functioning – efficient, competitive) and the 'construction of flanking mechanisms in civil society' (the 'outsourcing' to civil society groups like charities, NGOs and 'communities' of those interventionist functions which have been until late performed by the state) (Castree 2008a: 142). Thus, while neoliberal practices posit themselves as being apolitical and interested in economics alone, this assertion itself suggests that the realm of the economic is being totalised as the field of all human action – and thus, the field of economics must subsume any politics as well.

In this sense, Castree argues that the neoliberalisation of nature is a deeply political act, pushing for governance from the realm of economics. In trying to explain why neoliberalism is *necessarily* an environmental project, he uses the work of Karl Polanyi, James O'Connor, Ted Benton and Neil Smith and then identifies 'four broad rationales for nature's neoliberalisation on this basis' (Castree 2008a: 143). He argues that, if we see neoliberalism as a 'shell' in which capitalism pervades, then it offers the stakeholders of neoliberalism 'environmental fixes' to the unstable nature of sustained economic growth. First, it convinces one that regulation by the state will not do as much to protect nature as would privatising it. Second, it brings into the field of the market and finances those components of nature which were not subsumable under those terms of belonging. Third, neoliberalisation creates regimes where degrading the environment may be condoned for profiteering and materially subsumes such environment as can be degraded. Fourth, the state, which has to support capitalistic profiteering as well as to maintain its own legitimacy, can do two things in light of its own weakening role as state – it can outsource its regulatory action to private agencies, or it can maintain a 'minimal state' situation with little intervention. At the same time, to maintain its legitimacy, it can encourage its people to take (and feel) 'personal or communal responsibility for the "goods" and "bads" that arise from nature's neoliberalisation' (Castree 2008a: 149), using nationalism as a device in favour of neoliberalism.

In environmental conservation too neoliberalism has reared its ugly head (Igoe and Brockington 2007). This involves new ways of territorialisation of nature by partitioning of resources or whole regions in such a way that local communities are excluded. There is an emergence of networks between the states, NGOs and for-profit groups in the business of conservation guided by neoliberal ideologies that lead to changes in the lives of the people of those areas. For example, ecotourism is being encouraged to enhance economic growth and prosperity of the community as well as conservation of biodiversity. There is also an increase of corporate sponsorship of conservation organisations and management of protected areas by for-profit organisations. This is supported by the dismantling of state structures and practices that were limiting and what is offered in its place are simple solutions to complex ground realities. The systemic inequalities and relations of power that exist at the ground level are unaddressed while nature is sought to be conserved and protected through investment and consumption (Hartwick and Peet 2003). Territorialisation also highlights the commoditising of nature to meet the funding imperatives of neoliberalised states. These processes increasingly look at local people as having a flawed relationship with nature as well as the market. The environmental knowledge of the locals is devalued. Only a select few locals succeed in taking the opportunities offered by neoliberal conservation. The others experience the 'politics of disposability' (Giroux 2006, cited in Igoe and Brockington 2007: 444). The concept of citizenship itself is at stake since success is defined by ownership and access to transnational networks that define conservation.

Environmental injustice is reinforced on a world scale by green developmentalism (McAfee 1999). Transnational environmental institutions like the World Bank's environment section, patent offices of the OECD (Organisation for Economic Co-operation and Development) and so on, shift the burden of environmental costs to the erstwhile colonies that are rich in natural resources that according to them, do not know how to 'develop' and are under-polluted. The World Bank, for example, projects itself as the manager of the world's environmental resource flows and co-opts into neoliberalising development the challenges on the environmental front. Nature is projected as a 'globalised' resource, out of context of its social, spatial and historical background. This globalisation of environmental injustice is reflected in the recent recognition by green developmentalism of indigenous knowledge and traditional agricultural practices. However, their value is recognised only when they become a part of the global commodity circuits and not if they remain in local markets. Here the different 'values' of nature- subsistence value, exchange value or symbolic value are overtaken by commodity relations. In a globally unequal world, the purchasing power of developed countries reflects the existence of structural inequalities. The concerns of the northern countries for example in biological diversity conservation, global warming, ozone depletion and international waters and so on, reflect their priorities rather than the priorities of the southern nations, for example, land desertification. There is a devaluation of ecological and social relations in which biological diversity is embedded. Similarly, intellectual property rights to genetic resources are the basis of the biodiversity benefits under the Convention on Biological Diversity (CBD). This too is heavily in favour of developed countries of the global north and biotechnology companies. However, there have been resistance movements to green developmentalism and the movement of resistance to property in life. Here it is the 'indigenous people', the peasant movements in the global south who are developing coalitions with regional and international organisations and challenging green developmentalism. However, what is required is greater democracy and environmental justice and the need to question the agendas of the state and private agencies. This would help to unmask those who gain from these environmentally destructive policies and processes rather than blaming the destruction of biodiversity on an abstract market and policy failures (McAfee 1999).

We will now turn to the global south, particularly India, and examine the ways in which nature has been neoliberalised. But before that, a brief look at the term 'natural resources'.

Political economy of natural resources

'Natural resources' is a contradiction in terms. While 'natural' brings to mind the bounties of the earth like forests, minerals, water and so on that are not made by human beings, the term 'resources' refers to that which is culturally produced, exchanged and managed. In order to understand how natural resources have been imagined, appropriated and contested, Baviskar (2003) argues for the need to focus on the complex material and symbolic dimensions through the lens of cultural politics rather than political ecology alone.

One can trace the beginnings of the scientific management of nature and the development and commodification of nature to the colonial period. David Gilmartin (2003) in his essay *Water & Waste* examines the contradictory character of the construction of environment as 'resource' and as 'nature' and the related processes of development in colonial thinking in the large-scale canal irrigation works in the Indus Basin. In the colonial governments' agenda of increasing 'scientific' control over the environment as well as the control and management of indigenous communities, the term 'waste' and controlling of 'waste' was central. The aim of irrigation engineers was to tame and subdue nature and turn it into productive resource for a society of individual producers. While at the same time for the revenue administrators, the conversion of village commons or lands termed as 'waste' into productive resources for a community was defined on the basis of blood and genealogy rather than their roles in production. Through the explorations of the tensions between terms such as 'waste' and 'community', he examines the development of irrigation systems in the Punjab and the contradictions underlying the resource regimes of the colonial period.

Randeria (2007) highlights the shifts in regulatory, documentary and enforcement practices from the British colonial period to present times through the case study of the Gir forest. She analyses the processes of nature-making and state-building in the margins of the state, involving the interplay of state laws, the World Bank credit conditionalities and international norms of conservationist and human rights NGOs. More and more areas and communities in the peripheries have been brought under the control of a new regime of ecological 'development'. This globalised environmentalism is aided by the World Bank and international conservation NGOs like Worldwide Fund for Nature (WWF) (Goldman 2001: 503). Thus, standardised vocabularies for the organisation of nature, for example 'environment', 'natural resources', 'biodiversity' and 'protected areas', are created, leading to development of ecological surveillance and erasure of different experiences, systems of knowledge and survival strategies of the local communities. This, Randeria argues, is a political project inherited from colonial times and now has been exacerbated by the neoliberal environment policies of the World Bank.

The ways in which different communities relate to nature and the cultural forms of control and management that have emerged over time and space are always contentious and changing. We will now take a look at how this has played out in independent India, particularly in the post-reforms period.

Neoliberalising nature in India

The post–Second World War years until the early 1970s was a period of 'ecological innocence' (Guha 2000). During this time the erstwhile colonies like India were recovering from years of colonial rule and busy in developing and 'catching up' with the West. It was the period of state-induced development with emphasis on large dams, power projects and industrial agriculture wherein the costs to the environment and the people were overlooked. It was a period of continuous economic change and expansion. By the 1970s, however, the hope that the benefits of development would trickle down to the people at the lowest rung of the ladder had crashed. The reduction in forest cover, the increased amount of water pollution, the destruction of nature and the interdependent livelihoods led to a number of struggles by the marginalised groups whether they were fisherfolk, pastoralists, artisans or cultivators, especially the adivasis. This period witnessed a large number of struggles to protect the environment and the livelihoods of the local communities against the state, like for example the Chipko movement to protect the forests, the Silent Valley agitation for protection of biodiversity or the Narmada struggle against displacement. These popular movements emphasised the importance of local empowerment, the knowledge of the people and Gandhi's ideas of appropriate technology and rural development. It was against this background that a number of laws came into existence that regulated the forests, wildlife and water and air pollution in the years 1972–1974 and also led to the enactment of the Forest Conservation Act and the establishment of the Department of Environment in the year 1980. This was followed in 1986 by the Environment Protection Act and later in 1990 by the adoption of Joint Forest Management that encouraged community participation in forest management and questioned the control by the Forest Department.

The economy, which had declined in the 1980s, grew to a high in the post-reforms period of the 1990s. With the opening up of the economy, there was greater integration with global markets. This has led to an increase in the extraction industry, whether it is for biomass or the mining industry for coal, iron ore, bauxite and so on or the intensification of the use of ecological infrastructure like soil, water and air or the depletion of biodiversity. The role of the government has shifted from being the main investor to its withdrawal and instead it has taken on the role of being the main facilitator of investment. This shift in the goal of the state towards promotion of growth via indigenous private enterprise (rather than foreign capital) has been incremental (Kohli 2011, 2012). The growth in the Indian economy is accompanied by growing income inequalities, paralysis of pro-poor public policies, increase in capital intensity, growth of private industry and a decline in employment in manufacturing industry. The agenda has been to keep the large private players and multilateral agencies who are influenced by the Word Bank and financial institutions like the International Monetary Fund (IMF) happy, as they wield a lot of power over formulation of government policies. The burden of the rollback on public expenditure is borne by the poor. The rapid increase in wealth of the rich is aided by the transfer of land from the government to private corporations for a ‘public purpose’ such as mining, industrialisation and setting up of special economic zones (SEZs). In this process it is the Dalits, poor and marginalised groups that bear the burden of the economic progress of the rich (Bhaduri 2011). Since liberalisation, there has been a growing realisation of the tilt toward business houses in Indian politics. A shift that started in the 1980s has deepened over the last four decades to one that prioritises economic growth and business interests at present (Jafferlot et al. 2019). This has led to increased pressures on the environment and the unmaking of earlier systems of democratic decision making as green regulations were seen as a drawback on implementation of development projects that were to facilitate growth and development of the economy. With the victory of market forces, there is a push to quantify services provided by ecosystems like forests or mangroves. However, even though the government has put levies in place, it has not deterred the destruction of the environment or its restoration (Rangarajan et al. 2017). Instead it has led to the unfolding of green grabbing.

In pursuit of Lakshmi, we have forgotten how much a person needs to consume (Guha 2006). Demand is the new God and limits of natural resources as well as existing social systems are considered irrelevant. Within this capitalistic, managerial and scientific rationale it is assumed that for nature to be considered of any importance, it has to have a productive value attached to it. All other considerations of nature, including cultural, symbolic, religious or aesthetic, are omitted in favour of the idea of productivity for the market. What is considered of importance is how much we can produce or extract most cheaply. Since the opening up of the market, there has been a huge growth of infrastructure development like roads, highways, ports, airports, power stations and urban infrastructure that have led to the diversion of agricultural lands, forests and lands along coastal areas that has not only affected the environment but also a number of communities and their livelihoods.

The increasing demand for minerals has led to a spurt in the growth of the extraction industry by 75 per cent between 1993–1994 and 2008–2009. Mining for export is a major area for investment. A number of international mining companies like de Beers (South Africa), Vedanta (UK/India) and Alcan (Canada), to name a few, have invested in India (Shrivsatava and Kothari 2012; Kumar 2014). With the inflow of extractive capital in these resource rich areas, such as iron ore or bauxite mines in Orissa, Goa and Madhya Pradesh and the uranium mines of Jharkhand it is the tribal or other marginalised groups who are most adversely affected by the diversion of lands for mining. From self-reliant communities they are transformed into dependent, casual and brutalised labour in which women and children are the worst sufferers. Economic development has not necessarily translated into employment for these communities, belying the claims made for development. Besides the communities that are adversely affected, the environment is devastated and rendered barren due to mining. Forests and biodiversity of the area are destroyed and the wastes and debris generated by mining poison the water systems. Areas under wildlife sanctuaries and national parks are reorganised to support the extraction industry threatening the wildlife of the area. At the same time some sections of the powerful elite benefit with the inflow of extractive capital that generates a number of subsidiary avenues of accumulation in terms of land speculation, labour contracting, subcontracting, transportation and so on. This leads to deepening of extraction and exploitation and reinforces the power of extractive capital. The interests of the local elite, the political ruling class and extractive capital have come together and are now in control of the state apparatus. Any kind of resistance against extractive capital is seen as a challenge to the state and is strongly repressed (Kumar 2014).

Under globalisation, the stress on exports and monetary targets in terms of growth, rather than its impact on human welfare, has led to a boom in fisheries and aquaculture, floriculture, mining and commercial agriculture. The demands for marine products have risen tremendously in this phase and also the introduction of new technologies. Commercial scale intensive shrimp aquacultures have come up as it is mainly export-oriented. However, the social and environmental costs have been tremendous. The marine biodiversity has declined as now mechanised trawlers deplete the fish stock. This has led, as Castree argues, to the privatisation of the non-private, the marketisation of the non-market.

This destruction of the natural environment results in the multiple crises faced by peoples in India who are directly dependent on them for their survival and their livelihood. Even though some of these issues, such as food insecurity and shortages of water and fuel wood, were present even before globalisation in some form or other, they were to be alleviated by development and globalisation. However, the contrary seems to have occurred and in some cases the situation seems to have worsened. With globalisation, agriculture is in a state of crisis and it is unable to sustain livelihoods. Food security has worsened, with soil degradation and conversion of food production to cash crop production. Shortages in water are seen both in urban and rural areas for drinking purposes as well as for agricultural purposes. This is caused largely by mismanagement of water resources, pollution of water sources, lack of proper regulation of ground water usage and appropriation by private capital and the elite. All this has resulted in a crisis of livelihoods. The destruction of the land, water and natural resources has impacted communities of farmers, hunter-gatherers, fisherfolk, pastoralists, crafts persons and so on. Thus we see that development and especially globalisation has resulted in the state sponsored takeover by the private or corporate sector of agricultural lands, pastures, forests, water and other natural resources to the detriment of communities and the environment (Shrivsatava and Kothari 2012).

In this period of liberalisation, the environmental regulations and safeguards that had been brought into force after a lot of struggle by the people during the 1980s are being slowly and systematically unraveled due to opposition from industrialists and politicians. The Environmental Impact Assessment (EIA) notification of 1994, according to which clearances are mandatory for certain development projects, have been watered down. In 2006 a new environment clearance notification was brought out that made it much easier for industries and development projects to get clearance from the state without the involvement of any local community institutions. The entry of private players in areas of earlier exclusive control of the state has led to the amendment of laws in their favour. Liberalisation of the economy then has served the business interests in terms of providing natural resources and landscapes for development as well as consumption. Another way in which one can observe its impact is that conservation has become more business-oriented, with the global recognition of rights and responsibilities leading to an emphasis on ecosystem services that can then be commodified, prioritised and eventually monetised and regulated (Shahbuddin and Sivaramakrishnan 2019).

This can be seen starkly in the realm of forests and coastal areas. The diversion of forest lands for non-forest purposes has increased sharply in the globalisation phase. Demands for the scrapping of the Forest Conservation Act of 1980 have gained ground in this period. The coastal areas are also sites of contestation as they are being used not only for development of aquaculture and mechanised trawling, but are being increasingly viewed as sites for development of industries, tourism complexes and port hubs.

Hence the Coastal Regulation Zone (CRZ) notification of 1991 that provided some level of protection to the local fishing communities and to the mangroves and coral reefs began to be looked upon as a nuisance for industrial and commercial interests. In 2005–2006, modifications were suggested to the CRZ and a new Coastal Management Zone (CMZ) notification was proposed. Severe criticism by civil society groups and fisher communities for favouring industrial and commercial interests alone forced the government to allow it to lapse. However, there is rampant violation of the CRZ rules and a number of ports, tourism complexes and industries have come up along the coasts. The unregulated growth of tourism has had an immensely negative impact on the environment as well as the ways of life of local communities of the area. Another area of concern is the reorganisation and diversion of land within national parks and sanctuaries (Shrivastava and Kothari 2012).

In the last few years, in the pursuit of growth, the MOEFCC (Ministry of Environment, Forests, and Climate Change) has diluted environmental laws and procedures and made clearances easier. It has delinked the clearance required from National Board of Wildlife for forest clearance in and around national parks and sanctuaries. Procedures have been simplified under the Forest Conservation Act and requirements of public hearings have been done away with, allowing state governments rather than central government to give clearance in certain cases, there is a lack of non-government organisation participation, and the ban on industrial expansion in certain polluted areas has been lifted. Changes in important legislation like the National Green Tribunal set up in 2010 are being attempted. The Forest Rights Act that requires acceptance from the Grama Sabha for forest diversion is sought to be done away with as well as the progressive changes brought in the Land Acquisition Act of 2013. These are attempts to crack down on dissent and target civil society (Kothari 2014). One can thus see the material and discursive transformation of nature, which as Castree stated is an integral part of the neoliberal project. Thus, certain functions or services of nature are extracted and commoditised while a wide range of its qualities remain hidden and unrecognised. Through developments in biotechnology, commodities are created out of nature like genetic material and carbon molecules, which are patented and sold. Wild nature is also packaged and sold to private buyers. A valuation of nature is proposed by international banks, lending groups as well as recipient governments. Thus, the Natural Capital Framework hopes to weed out the wasteful users of nature through high pricing of nature that the poor can ill afford. In World Bank terminology, all the users are stakeholders who are forced into a level playing field. Here the same resource is used by those who can pay a higher price than by those who use it for local, community-level needs that are low in economic value. Since the 1990s at the global level, 'payment for environmental services' (PES) schemes were started to provide subsidies or financial incentives to protect the environment. In India, the Biological Diversity Act (2002) made according to the Convention on Biological Diversity (CBD) promises 'benefit sharing', for example, monetary compensation given for sharing local indigenous knowledge as well as access to biodiversity with pharmaceutical companies or research organisations. This is set to change the very meaning of biodiversity and benefits. Furthermore, property rights or ownership rights in nature have been developed to encourage protection of investments. This has also helped environmental movements to negotiate with extractive capital for better compensations or trying for patents before companies do so and thus not being only the victims of development. In a neoliberal state, this compartmentalisation of land, water and forests into individual ownership or use rights has led to the transformation of the idea of government property held in trusteeship for the community and the idea of multiple claims of the commons to the idea of nature as private property that supports large industrial users. This is supported by environmental regulations, policies and laws that convert land, water, forests and animal and plant species into commodities that have been assigned an exchange value. For the global economy, nature is but a subset (Kohli and Menon 2016).

We will now try to see how the various sections and chapters of this book interlink with the core ideas of this Introduction.

Sections of the book

The first section is titled 'Neoliberal Governance, Environment and Gender' with chapters laying out the theoretical backdrop and examining the intersections of nature, nation and science through a critical feminist lens. In the chapter titled 'Cutting the Gordian Knot: Environmentalism, Capitalism and the Metabolic Rift' Sudha Vasan develops a theoretical understanding of the source of the current ecological crisis. The chapter draws out the contradictions in the relationship between capitalism as a mode of production and the contemporary efforts to deal with the ecological crisis. Discussing the problems within capitalism, the chapter then focuses on the problem of perception of the ecological crisis in the contemporary unequal world. Vasan argues that due to the inability to understand the crisis, the solutions that emerge are reductive and tend to shift the problem rather than resolve it. Finally, she argues that environmentalism needs to be understood based on ones understanding of the problem rather than on the social location of its members.

Linked to this understanding of the ecological crisis and exploring it in the Indian context is Gita Chadha's chapter titled 'Nature, Nation, Science and Gender'. In this chapter she explores the intersections of nature, nation and science through a critical feminist lens. She argues that modern, Western science has been critical to the project of development and progress that envisioned the nation as the mother in the nationalist discourse and feminised it through multiple discourses. With the development of science studies, science has been critiqued by feminist, post-colonial, peace and environmental movements. Analysing the work of J. C. Bose, Chadha tries to understand if it is an alternative approach to the ideas of mainstream science. Further, she looks at the way feminists have critiqued the feminification of the nation through gender symbolism and the visual in India. In exploring these intersections in the Indian context, Chadha argues for the development of new feminist insights in studies of nation and nationalism, for science studies and for environmental studies that are plural, inclusive and compassionate. Chadha introduces a new conceptualisation of nature through a feminist understanding of science that has reconfigured ideas, values and practices around nature.

Chandan Kumar Sharma explores further the idea of nature and nation through his chapter titled 'Building "India's Future Powerhouse" '. He highlights the issues of North East India through his critique of the government of India's policies to turn the region into a powerhouse through the generation of hydel power from its many perennial rivers. He analyses the shifts in the discourses of development for the North East region from a 'security paradigm' to a 'development paradigm' with the amount of expenditure on infrastructure in the post neoliberalisation phase of the 1990s with public and private investments. This he argues is to cater to the increased electricity needs of industries and cities of 'mainland' India as well as to the Indian state's need to 'nationalise space' in order to assert its presence in the region, especially in Arunachal Pradesh to counter the Chinese territorial claims. Popular resistance to the projects that would affect not only the river ecosystems but also the livelihoods and the cultural heritage of the people of the region has arisen. What is required is development that is in sync with the social, cultural, economic and ecological specificities of the region in consultation with civil society. The idea of green developmentalism argued by McAfee is reflected in Sharma's chapter with a regional twist to it.

The chapters in the next section, titled 'Community Politics and Livelihoods', focuses on the micro level of the community. First, Ritambhara Hebbar reflects on her field work in the Nilgiri biosphere reserve and the politics of conservation. Her chapter is followed by D. Parthasarathy and Hemantkumar A. Chouhan on the new coastal claims and the impact of neoliberalisation on the fisher community in an urban setting. The next chapter in this sub theme is by Hita Unnikrishnan, B. Manjunatha and Harini Nagendra, on lakes as urban commons.

Ritambhara Hebbar in her chapter 'Nilgiri Biosphere Reserve: Reflections from the Field' points to the complete lack of concern for the tribal communities of the region in environment conservation and policy reports, particularly the Nilgiri Biosphere Reserve (NBR) project. Through a reading of official documents pertaining to the NBR project she points out how environmental projects have reproduced the larger politics of tribal assimilation that dominate the tribal question in India. Conservation politics start from the assumption that managerial fixes and technical knowhow are superior and can be reproduced across sites, overlooking the local history and political economy of the local tribal communities. Through interviews with the local tribal communities and activists, Hebbar highlights the nature of governance in the NBR and the multiple ways in which the tribal communities have been marginalised and denied claims to the forest. She argues that the conservation projects are modes of governance that introduce newer institutional practices to control and manage resources rather than for conservation of biodiversity of the region. Her work points to the way environmental agendas submerged the voices of tribal peasants and overlooked the core issues facing the tribes of the region. Hebbar's chapter reflects the concerns put forward by Igoe and Brockington (2007) in terms of the need to look through the neoliberal lens to understand the politics of conservation.

In their chapter titled 'New Coastal Claims and Socio-Legal Contestations in Mumbai: Artisanal Fishers and the Problematic of the Urban Environment', D. Parthasarathy and Hemantkumar A. Chouhan explore the marginalisation of artisanal fishers due to the new coastal claims and encroachment in the CRZ-notified areas of the metropolitan region of Mumbai. They argue that the degradation of coastal areas and the livelihoods of the fishing communities are compounded by neoliberal urbanisation and speculative capital, which are making new claims and demands on the urban coastal zones. Besides this are the conservation interests and development projects in the region. The fisher communities caught in this double jeopardy are fighting back to defend their community rights and restore marine ecologies through formal and informal strategies. There is a need for an alternative understanding of social inequalities and legal transformations with regard to the coastal commons, especially since both the state and market forces are redefining the environment to increase control and help in the process of accumulation by dispossession. The authors analyse the contested claims through the lens of political ecology and environmental justice and argue for the need to situate environmental law and governance as a 'semi-autonomous social field' that needs to be studied from that perspective as well.

Hita Unnikrishnan, B. Manjunatha and Harini Nagendra in their chapter titled 'Bonds That Divide: Urbanisation and Erosion of the Commons' explore the impact of transformation in the ecology and character of the lakes on the local communities dependent on these resources in the southern city of Bengaluru. They argue that the lakes are dynamic spaces that provide to the communities dependent on them recreational, provisioning and cultural services. However, contemporary planning measures have tended to exclude communities accessing ecosystem services provided by the lake. The neoliberal regimes, they argue, have threatened and transformed the urban commons. It has also created systems that exclude communities that used to rely on the lakes for their sustenance and livelihood requirements from its stewardship and is in the hands of mainly migrant populations with no real connection with the lake.

The next section, titled 'Marketisation and the Environment' examines the processes of marketisation of the environment in subtle ways. Arun de Souza's chapter looks at the neoliberalisation of watershed policies in Maharashtra, followed by Seema Kulkarni's chapter on privatisation of canal irrigation.

In the chapter titled 'Playing with Coloured Spectacles: Neoliberal Witchcraft as Played Out through Watershed Policies', Arun de Souza outlines the government's policy on watershed management and argues that it is not just about conserving nature and developing sustainable livelihoods but also about the transformation of its participants into neoliberal subjects ostensibly through processes of democratisation. Through neoliberal witchcraft, he points to the transformation of the landless local communities into encroachers and trespassers on their own customary landholdings. He then points to the use of science and technology to transform the primitive locals into modern citizens of a developmentalist neoliberal state. Through these processes the state exercises its power over hitherto excluded village communities in the borderlands and transforms them into economic citizens within the grid of global capitalism. He attempts to show how a capitalistic, managerial and scientific rationale is used which presumes that nature in order to be valued must have a productive value. However, he points to the micro politics of power at play at the ground level where he highlights how people manage to subvert and resist these processes even as they acquiesce in the larger development plan. Through his study of watershed management in Maharashtra he discusses the contestation, subversion and co-optation between the dominant Maratha caste and the marginal communities.

In her chapter on 'Canal Commands and Rising Inequity' Seema Kulkarni highlights the increasing privatisation of a public resource. She analyses how public sector irrigation projects in Maharashtra are increasingly being used to harness groundwater. This has been a gradual ongoing process which has shifted benefits from public to private. Changing discourse in the water policy prioritising private over public interests is contributing substantially to these inequities. This she argues, reflects the concern that the neoliberal language has become dominant in steering water policies and programmes that hold implications for water access, allocations and decision making.

In the final section, 'Law, Politics and Resistance', Arupjyoti Saikia looks at the evolution of Kaziranga National Park in the context of contested ecological histories of Assam. This is followed by Geetanjoy Sahu's chapter on the legal terrains of ecological jurisprudence. The final chapter by John Kurien tries to look for alternatives in the lives and life worlds of the common fisher peoples.

In his chapter titled 'Rhinoceros in Kaziranga National Park: Nature and Politics in Modern Assam' Arupjyoti Saikia tries to understand the historical evolution of the park and its political fallout in the context of the complex ecological histories of Assam. He argues that the park has become the site of increasingly intense debates surrounding the questions related to nature's future, law and rights. This debate to decide the future of the park is influenced by questions relating to the flagship one-horned rhino and the agrarian communities of the region. However, he argues that the debates around the park are immune from the complex ecological, historical and political realities of Assam and the environmental past and the present. He traces the history of hunting in the floodplains of the Brahmaputra valley and the rapid rise of hunting the rhino in colonial India. In post-independent India, he traces the efforts to save the rhino and its linkages with the making of Assam's cultural heritage. He argues for the need to look at the competing rights of the local communities, the land and the animal while trying to understand the formation of the natural space of Kaziranga.

In the chapter titled 'Environment Movements and the Indian Supreme Court', Geetanjoy Sahu examines the context of the emerging legal and normative perspectives on environmental law and governance. Through a review of a series of environmental judgments delivered by the Indian Supreme Court, Sahu traces the differing environmental values that are given priority by the judiciary and its impact on the environmental movements in India. The different strands of the green discourse in India range from a strong material interest in the environment in relation to livelihood needs, to the non-material interests that relate to the intrinsic value of nature, to the middle-class concerns for a healthy environment. However, the recognition and legitimation of one particular environmental value over others depends, Sahu argues, on the prevailing political economy, the political regime and the strategies adopted. Reviewing some of the important environmental judgments delivered over a 30-year period from 1980 to 2010, he traces the shifts from need for a healthy environment to prioritising large infrastructure projects ignoring environment and human rights concerns. He emphasises the need to use multiple strategies to address environmental issues and use the judicial path only as the last resort.

In the chapter titled 'Wise Sayings from an "Ecosystem" Community: Reflections from a Search for Challenging Neoliberal Worldviews on Nature', John Kurien reflects on the words of ordinary fisherfolk from around the world whose livelihoods and worlds are intimately tied to the dynamic aquatic ecosystem. The words of these ordinary fisherfolk have reframed the way resources and processes in nature and the interactions between them are perceived. He argues that in today's world we are tutored to look only at the material aspects of nature and treat it as a source of resources and a sink for wastes. The perspective offered by the fisherfolk challenges this mindset and provides insights for an agenda of resistance from total co-option into the neoliberal worldview. He points out that only ecosystem people depend on nature for their life and livelihood, unlike biosphere people. Their connectedness with Mother Sea and their dependence on it gives them complete faith in nature's bounty even in the face of despair and destruction. Hence, they have the courage to resist the use of inappropriate technologies.

Conclusion

Around the world, new ways of compartmentalising, commodifying and privatising nature are increasingly leading to a new global discourse based on 'green' market economics and the use and repair of environments. In the past too there have been interventions in the name of environment protection, however, in the present times, green grabbing operates through novel legal and market mechanisms. New ways of valuing and appropriating nature are being legitimised. In this process of 'green grabbing' a new set of actors, processes and resistances have come into play that seem to suggest new ways of understanding them, both methodologically as well as analytically. The ongoing struggles and mobilisations at the grassroots have built networks not only between those directly affected by displacement but also between peasants and farmers who are also affected by the environmental disaster that is unfolding in these areas as well as middle-class, urban-based environmentalists. In these resistance movements against global extractive capital and the neoliberal state in the global south, one can see the convergence of red (social justice) and green (ecological) issues.

This neoliberalisation of nature and green grabbing that is associated with it is context specific. As Moore argues, 'capitalism is not an economic system; it is not a social system; it is *a way of organizing nature*' (2015: 13). Hence, what is needed is specific ways in which one can take back nature from the logics of market control. This would require a reframing and recasting of the debate on the environment in novel ways. Only critiquing the processes of neoliberalisation of nature is inadequate. What is required is to keep one's ear to the ground and find alternatives that take into account the complexities of the ground reality and the interconnections between the environment and human societies. As articulated by the set of chapters in this volume, the responses to the neoliberalisation of nature are varied and are socially and politically embedded in the context. They respond in multiple ways to the appropriation of nature- and market-oriented policies of the government and have their own ways of responding in uniquely transformative ways. To conclude, one wonders if there could be an ecological turn in academic scholarship given the fact that the environmental issue is of central concern.

Notes

1 A section of this chapter was originally published as Manisha Rao, 'Reframing the Environment in Neoliberal India: Introduction to the Theme', *Sociological Bulletin*, Vol. 67, Issue 3, pp. 259–274. Copyright 2018 © Indian Sociological Society. All rights reserved. Reproduced with the permission of the copyright holders and the publishers, SAGE Publications India Pvt. Ltd, New Delhi.
2 The national conference 'Reframing the Environment: Resources, Risk and Resistance in Neoliberal India' was conceptualised and coordinated by me and organised at the Department of Sociology, University of Mumbai with partial financial support from ICSSR New Delhi, on 21–22 January 2016.
3 The concept of environmentality is used by Arun Agrawal to describe a disciplinary form of conservation governmentality. In his study of the communities in Kumaon, he analyses the new ways of thinking and acting in relation to the environment and the constitution of 'environmental subjects'.

References

Agrawal, A. 2005. *Environmentality: Technologies of Government and the Making of Subjects* (New Ecologies for the Twenty-First Century). Durham, NC: Duke University Press.

Arsel, M. and B. Buscher. 2012. Forum issue, with a debate section on 'neoliberal market mechanisms in environmental and conservation policies'. *Development & Change*, 43(1).

Bakker, K. 2005. Neoliberalizing nature? Market environmentalism in water supply in England and Wales. *Annals of the Association of American Geographers*, 95(3), 542–565.

Bakker, K. 2009. Neoliberal nature, ecological fixes, and the pitfalls of comparative research. *Environment and Planning A*, 41(8), 1781–1787.

Baviskar, A. 2003. For a cultural politics of natural resources. *Economic and Political Weekly*, 38/48 (2003): 5051–5055.

Bhaduri, A. 2011. Predatory growth. In P. Balakrishnanan, ed. *Economic Reforms and Growth in India: Essays in Economic and Political Weekly*. Hyderabad: Orient Blackswan.

Brockington, D. and R. Duffy. 2010. Capitalism and conservation: The production and reproduction of biodiversity conservation. *Antipode*, 42(3), 469–484.

Brockington, D., R. Duffy and J. Igoe. 2008. *Nature Unbound: Conservation, Capitalism and the Future of Protected Areas*. London: Earthscan.

Castree, N. 2005. The epistemology of particulars: Human geography, case studies and 'context'. *Geoforum*, 36(5), September, 541–544.

Castree, N. 2008a. Neoliberalising nature I: The logics of de- and re-regulation. *Environment and Planning A*, 40(1), 131–152.

Castree, N. 2008b. Neo-liberalising nature II: Processes, outcomes and effects. *Environment and Planning A*, 40(1), 153–173.

Castree, N. 2011. Neoliberalism and the biophysical environment 3: Putting theory into practice. *Geography Compass*, 5(1), 35–49.

Fairhead, J., M. Leach and I. Scoones. 2012. Green grabbing: A new appropriation of nature? *The Journal of Peasant Studies*, 39(2), April, 237–261.

Gilmartin, D. 2003. Water and waste: Nature, productivity and colonialism in the Indus basin. *Economic and Political Weekly*, 38(48), November 29–December 5, 5057–5065.

Giroux, H. 2006. *Stormy Weather: Katrina and the Politics of Disposability*. Boulder, CO: Paradigm Publishers, cited in Igoe and Brockington 2007: 444.

Goldman, M. 2001. Constructing an environmental state: Eco-governmentality and other transnational practices of a 'green' World Bank. *Social Problems*, 48(4), 499–523.

Goldman, M. 2005. *Imperial Nature: The World Bank and Struggles for Social Justice in the Age of Globalisation*. Yale University. Hyderabad: Orient Longman.

Guha, R. 2000. *Environmentalism: A Global History*. New Delhi: Oxford University Press.

Guha, R. 2006. *How Much Should a Person Consume? Thinking Through the Environment*. Ranikhet: Permanent Black.

Hartwick, E. and R. Peet. 2003. Neoliberalism and nature: The case of the WTO. *The Annals of the American Academy of Political and Social Science, Rethinking Sustainable Development*, 590, November, 188–211.

Harvey, D. 2003. *The New Imperialism*. Oxford: Oxford University Press.

Harvey, D. 2005. *A Brief History of Neoliberalism*. Oxford: Oxford University Press.

Heynen, N., J. McCarthy, S. Prudham and P. Robbins (eds.). 2007. *Neoliberal Environments*. London: Routledge.

Heynen, N. and P. Robbins. 2005. The neoliberalisation of nature: Governance, privatisation, enclosure and valuation. *Capitalism Nature Socialism*, 16(1), 1–4.

Igoe, J. and D. Brockington. 2007. Neoliberal conservation: A brief overview. *Conservation and Society*, 5(4), 432–449.

Jafferlot, C., A. Kohli and K. Murali (eds.). 2019. *Business and Politics in India*. New York: Oxford University Press.

Kohli, A. 2011. Politics of economic growth in India, 1980–2005: The 1990s and beyond. In P. Balakrishnanan, ed. *Economic Reforms & Growth in India-Essays in Economic and Political Weekly*. Hyderabad: Orient Blackswan.

Kohli, A. 2012. *Poverty Amid Plenty in the New India*. Cambridge: Cambridge University Press.

Kohli, K. and M. Menon (eds.). 2016. *Business Interests and the Environmental Crisis*. New Delhi: Sage.

Kothari, A. 2014. A hundred days closer to ecological and social suicide. *Economic & Political Weekly*, xlix(39), September 27.

Kumar, K. 2014. Confronting extractive capital: Social and environmental movements in Odisha. *Economic & Political Weekly*, xlix(14), April 5.

Leach, M., J. Fairhead and J. Fraser. 2012. Green grabs and biochar: Revaluing African soils and farming in the new carbon economy. *Journal of Peasant Studies*, 39(2), 285–307.

McAfee, K. 1999. Selling nature to save it? Biodiversity and the rise of green developmentalism. *Environment and Planning D: Society and Space*, 17(2), 133–154.

McAfee, K. 2011. *Selling nature to finance development? The contradictory logic of 'global' environmental-services markets*. Paper presented at the conference on 'NatureTM Inc? Questioning the Market Panacea in Environmental Policy and Conservation', Institute of Social Studies, The Hague, 30 June–2 July.

McCarthy, J. and S. Prudham. 2004. Neoliberal nature and the nature of neoliberalism. *Geoforum*, 35, 275–283. McCarthy and Prudham 2004, cited on p. 139.

Moore, J. 2015. *Capitalism in the Web of Life Ecology and the Accumulation of Capital*. London: Verso.

Prudham, S. 2009. Commodification. In N. Castree, D. Demeritt, D. Liverman and B. Rhoads, eds. *A Companion to Environmental Geography* (pp. 123–142). Chichester: Blackwell.

Randeria, S. 2007. Global designs and local lifeworlds: Colonial legacies of conservation, disenfranchisement and environmental governance in postcolonial India. *Interventions*, 9(1), 12–30. ISSN: 1369-801X print/1469-929X online. Taylor & Francis.

Rangarajan, M., R. Sarkar and R. Agarwal. 2017. *The Problem*. Nature's present – A symposium on understanding conflicts around nature in contemporary India, Seminar No. #690, New Delhi, February.

Shahbuddin, G. and K. Sivaramakrishnan (eds.). 2019. *Nature Conservation in the New Economy-People, Wildlife and the Law in India*. Hyderabad: Orient Blackswan.

Shrivastava, A. and A. Kothari. 2012, 2014. *Churning the Earth: The Making of Global India*. New Delhi: Penguin Books.

Vidal, J. 2008. The great green land grab. *The Guardian*, 13 February. www.theguardian.com/environment/2008/feb/13/conservation

Part I

NEOLIBERAL GOVERNANCE, ENVIRONMENT AND GENDER

2

CUTTING THE GORDIAN KNOT

Environmentalism, capitalism and the metabolic rift

Sudha Vasan[1]

> Definition of *Gordian Knot*
>
> 1: an intricate problem *especially*: a problem insoluble in its own terms – often used in the phrase *cut the Gordian knot*
>
> 2: a knot tied by Gordius, king of Phrygia, held to be capable of being untied only by the future ruler of Asia, and cut by Alexander the Great with his sword
>
> Merriam-Webster Dictionary

The Anthropocene is a popular term used to capture the distinct impact of humanity on the earth at a planetary geological scale (Steffan et al. 2007). Cascading changes in the earth's bio-geo processes suggest a threat to the conditions that sustain human life on earth; scientists have identified nine major processes that are crucial to maintaining human life on earth, the *planetary boundaries* (Rockstrom et al. 2009): climate change, ocean acidification, stratospheric ozone depletion, the nitrogen and phosphorus cycles, global fresh water use, change in land use, biodiversity loss, atmospheric aerosol loading and chemical pollution. The ecological crisis we face today is the potential destruction of the conditions that sustain human life on earth.

A diverse range of social efforts – environmentalisms – has emerged across the world as a response to the gravity of this situation. From grassroots movements in the global south that emphasise sustainable livelihoods to post-industrialist de-growth movements, from calls to massive collective mobilisation to individual moral choices, from radical demands to halt growth to greening and growing the existing economy, the range and diversity of environmental movements across the world is truly astounding. The need for multiple and varied efforts to resolve such a fundamental crisis facing humanity is widely recognised. How do we understand these varied responses and their impulses and impacts on the ecological crisis?

One of the most productive and recognised categorisations of environmentalism has emerged from the body of work by Ramachandra Guha (1988, 2000) and Joan Martinez Alier (2002), both individually and together (see Guha and Martinez-Alier 1997, for example). Three currents of environmentalism are summarised in Martinez-Alier (2002) as the 'cult of wilderness', the gospel of eco-efficiency' and the 'environmentalism of the poor'. The distinction between post-industrial environmental concerns of the North and the livelihood focused environmental concerns of the South is also a legacy of these works. In an early essay, Guha (1988) also distinguishes environmental movements in India on their ideological bases, and he finds three strands: crusading Gandhians, appropriate technology and ecological Marxists. Others have adapted this framework to different contexts and developed categorisations such as romantic environmentalism, environmental management and environmental justice, each placing issues in specific interpretations of past, present and future (Schoenfeld 2005). Other scholars have used specific categorisations, such as bourgeois environmentalism (Baviskar 2003). While each of these categorisations of environmentalism have their particular relevance, this chapter underlines the importance of a focus on the understanding of the problem when classifying a social movement, since this understanding centrally contributes to the ability to handle the problem. Such categorisation also helps to avoid essentialisation and romaticisation that a focus on social location may engender.

At the core of the crisis and therefore its solution, the first section of this chapter argues, is understanding of the human relationship to non-human nature. Each form of environmentalism has at its core a stated, unstated or even unrecognised assumption about the human-nature relationship. Drawing on Foster's work (1999, 2000), we highlight the distinctiveness of a dialectical-materialist understanding of human-nature and situate it in the debate around the fundamental nature of capitalism and its potential (or lack thereof) for resolving the ecological crisis. The regime of capital is critiqued as carrying the seeds of destruction within itself, creating a metabolic rift and being inherently a globally destructive force (Foster et al. 2011). Drawing from this literature, this chapter will draw out the contradictions in the relationship between capitalism as a mode of production and our contemporary efforts to deal with ecological crisis.

The first section summarises theoretical developments in understanding the source of our current ecological crisis. The historical establishment of a metabolic rift and the shifts engendered as solutions to this problem within capitalism are discussed. The second section focuses on the problem of perception of the ecological crisis in the contemporary world. An unequal world cannot be a sustainable world as standpoint influences even the perception of an impending precipice and consequently any form of collective action. Given this inability to understand the crisis, the solutions that emerge are reductive and tend to spatially, temporally or socially shift the problem rather than resolve it. Therefore, the final section argues that environmentalism – or social movements to protect the environment – needs to be categorized and understood based on the movements' understanding of the problem rather than on the social location of its members.

The metabolic rift

Foster's (1999) work highlighted Marx's prescient understanding of the human-nature relationship as a metabolic relationship where there is a constant flow and recycling of energy and matter. Humans as a part of nature also constantly transform it while transforming themselves. However, the emergence of capitalism generated a rift that disrupts this smooth metabolism. Marx draws from the work of German chemist Justus Freiherr von Liebig to elaborate on the metabolic rift that is produced by the separation of town and country engendered by capitalism. With the first wave of industrialisation, for the first time, the agricultural output consumed by human populations is not recycled back into the field. The food produced from the soil in the country is transported to the town for the consumption of workers, while the waste generated by this mass of humans is dumped in the cemented town where it has no way of recycling back into the agricultural soil. The link between what is extracted from the soil and what is recycled back into it is thus broken, resulting in the accumulation of waste in the town and the depletion of nutrients in the soil in the country. Thus, a rift is created in the cycle of nutrients from the soil into human bodies and back into the soil. This rift is not generated by agriculture per say, but by the separation of town and country that underlies capitalist organisation of production. This is the original metabolic rift that capitalism introduced in the metabolic relation of humans to nature.

Marx goes on to describe how capitalism deals with this metabolic rift. As the soil is robbed of its nutrients by the segregation of town and country, nutrients now have to be added to it by transporting it from other locations. In the imperialist phase of capitalism, nutrients are imported from the colonies, using war, colonisation and slave labour. In the 19th century, Clark and Foster (2009) describe how China, Peru, Chile, Britain and the United States were linked through a global metabolic rift. Armies and wars were put to work to bring manure from distant colonies. The problem of British agriculture was 'solved' through guano imperialism, where entire islands were colonised and slave labour used to remove and transport bird shit across continents. The problem of loss of soil fertility was thus shifted to the colonies through colonial depredations. Therefore, "ecological imperialism has meant that the worst forms of ecological destruction in terms of pillage of resources and disruptions of sustainable relations to the earth, fall on the periphery rather than the center" (ibid.: 330).

The development of chemical fertilizers marked the end of guano imperialism, when commercially produced fertilizers replaced the nutrients stolen from the soil through capitalist agriculture. While agriculture itself is a process in which the nutrients of the soil are transformed into food that is consumed by people for their sustenance, in the metabolic cycle between humans and nature, human waste is recycled back into the soil. Capitalist production systems however sever this metabolic cycle by concentrating huge populations in industrial centres disconnected with the agrarian countryside. Converted soil nutrients are transported great distances and the waste produced accumulates in urban centres and is washed out to the sea rather than cycling back to enrich the soil. While urban waste creates its own problems in urban metabolism, agriculture relies on other means to replenish the soil robbed of its nutrients and problem is temporarily shifted.

Over the next century, excessive fertiliser use in intensive capitalist agriculture resulted in further ecological problems including extinction of pollinators, mutation of crop pests and poisoning of soil and water that contaminated the food chain. In response to each of these problems, technological market solutions shift the problem either spatially or temporally, resulting in other problems.

Examining contemporary global development and economic flows, several scholars have revealed the material-ecological flows that transform ecological relations between town and country or across global regions (Burkett 1999; Clark and Foster 2009; Hornborg 2003). This understanding of the ecological crisis provides particular insights into the nature of the current ecological crisis. It demands an examination of our fundamental relationship with nature and the processes that affect this relationship. It also centrally implicates the organisation of production and reproduction that undergirds our relationship to nature. An understanding of the metabolic rift underlines the depth of the problem that we refer to here as the ecological crisis, and the limitation of our current efforts to deal with the ecological crisis. The problem is not merely the perceptible changes in temperature or sea level rise. These are critical symptoms. The deeper problem is a rift in a most basic human-nature relationship. Symptomatic redressals only tend to shift the problem spatially, temporally or socially, and fail to transform the conditions that create the crisis.

Standpoint and responses to the ecological crisis

Climate change is perhaps the best-known planetary boundary, and one of the two where many scientists claim the tipping point has been crossed. Rise in global temperatures and increase in the frequency of extreme weather events have been recorded consistently enough to convince most sceptics to be at least concerned. While climate change denialists today are fewer in number, it is significant that it is often an opinion held by the most powerful and most consumptive societies and individuals, who contribute significantly to climate change.

A recognition of 'standpoint' drawing from feminist theory (Jaggar 1983; Harding 1986) allows us to understand the complexity of dealing with planetary boundaries in an unequal world. The erosion of the conditions of human species existence is an indivisible common threat that affects humanity as a whole. However, the world's most disadvantaged people are saddled with the burden of environmental degradation disproportionately: they have contributed the least to the damages so far since they have not been recipients of the benefits of development, they have suffered most from the damages caused whether it is climate change or loss of forests or toxic pollution and they are forced to bear a heavier share of the considerable costs of the clean up or preservation agendas while having the least opportunity to influence policies on these issues. For instance, growing evidence suggests that the impacts of observed and future climate change are and will be spatially and socially differentiated (Shue 1999; Adger 2006). The worst effects of climate change will fall disproportionately on those living in sub-Saharan Africa, small islands in the Pacific and Indian oceans and deltaic regions of South and Southeast Asia, Egypt and China. (IPCC 2007). Climate change is expected to hit developing regions the hardest. Its effects such as higher temperatures, changes in precipitation patterns, rising sea levels and frequent weather-related disasters will pose a risk to food and water supplies. Evidence shows that not only are the poorest people often more exposed to specific climate change impacts, they are also more vulnerable to those impacts and find it harder to recover when they occur (Adger 2006). Climate change will widen existing inequalities, globally and locally. The extent of vulnerability will depend on more than just terrain and climatic conditions: "the fraction of the population living in low-lying regions, the area and proportion of the country inundated, its wealth and economic conditions, and its prevailing political institutions and infrastructure will all be of relevance" (Byravan and Rajan 2010: 240).

Countries with more control over resources may be able to better adapt to changes. Even within developed countries, studies have shown that climate change will disproportionately affect the poorest in society (Wolstenholme 2009). Her research shows that the people who are likely to be most vulnerable to the impacts of climate change are those living in places at risk, people who are already deprived by the health, level of income, the quality of their homes and mobility; as well as people who lack awareness of the risks of climate change and the capacity to adapt, and who are less well supported by family, friends and state and non-state agencies. The most disadvantaged people suffer the most from environmental degradation, including in their immediate personal environment, and disproportionately lack political power to do anything about it (UNDP 2011). No one, however resourceful or powerful will remain unaffected by the planet-level transformations that are happening. But the gradual and often invisibilised nature of climate change or environmental transformations allows differential appreciation of the crisis. So the urgency of the problem is differentially experienced. This differential experience, apart from questions of justice and ethical responsibility, raises the challenge of differential perception and understanding of the nature of the ecological crisis. Responses to climate change reflect these material subjectivities.

Another explanation for the fragmented appreciation of the severity of the current crisis relates to differential expectations for the future based on the experiences and expectations of the present. Consider the concept of sustainable development, drawing from the most quoted definition from the Brundtland Commission (WCED 1987): sustainable development is development that meets the needs of the present without compromising the ability of future generations to meet their own needs. Sustainable development emphasises intergenerational equity, which is a very important ideal. But this needs to be moderated with the understanding that not every section of humanity would be happy with a simple reproduction of the society we have today. 'Meeting the needs of today' is a complex social statement where distribution of resources is not according to needs, but according to ability to control resources. In a world divided by class, race, caste, gender and other inequalities, the continuation of current conditions for future generations is an inadequate goal. Widespread acceptance of the intergenerational equity ethic then presupposes dealing with intra-generational equity.

Thirdly, there is a fundamental difference in the way the nature of the problem is perceived, again arising from differential experience of environmental change. In regions and sections of the population of the world with higher levels of 'development' and consumption, the problem is recognised as an environmental crisis, as a threat to Earth or ecology. There is a conscious environmental movement therefore that seeks to protect the environment. In contrast to this, among vast sections of people in less-developed regions of the world, where clearly the fallout of ecological change is more current and immediate, the problem is perceived as a human crisis. The problem is then loss of human lives and livelihoods, and what many social movements seek to protect is human life and livelihood rather than an abstract environment. The former then sets out to protect the environment while the latter sets out to protect human life, both often obstinately refusing to recognise the interconnectedness of the issue. In the latter, this recognition has seen a steady growth, with an ecological narrative becoming part of social movement narratives to various extents. Social movements of the latter variety, which are often struggles over control of natural and livelihood resources, have sometimes been referred to in the literature as environmentalism of the poor (Guha and Martinez-Alier 1997). While the literature so far has rightly focused on the strengths of recognising the contribution of environmentalism of the poor for a more sustainable future, it is also significant that that the definition of the problem and consequently the solutions proposed by these movements is not always easily adjusted with the post-industrial environmental ethics and movements.

Denial of the crisis is most prevalent among the most affluent and advantaged people and nations. Apart from outright denial, other responses that dominate include attempts to isolate and shift the problem across sectors and regions, and forcing some sections to bear the cost of cleaning up the mess or enforcing conservation at 'distant' places through domination or market mechanisms.

However, environmentalism needs to be recognised as an important ideological trend that has a fairly long history that matures in the late Anthropocene. Environmentalism, that recognises the relationship of nature and human society has a variety of trajectories and forms. One of the more popular classifications of environmentalism has been to recognise the divide between the post-industrial environmentalism of the North and the environmentalism of the South among poorer or marginal peoples/regions/nations (Guha and Martinez-Alier 1997; Martinez-Alier 2003). This distinction is based on social location of the actors, forms of expression of concerns regarding the environmental crisis and the resolutions that are imagined and demanded. This categorisation emphasises actor identities to categorise how and why people organise to deal with the global ecological crisis. This has been particularly useful in focusing attention on inequality and has been central within environmental justice movements. However, this classification does not throw much light on the potential of either movement to address (or resolve) the ecological crisis that we are living in today.

Real and false gods

The metabolic rift is a direct indictment of the central logic of capital: Grow or Die (GOD). Capitalism as a system cannot survive without expanded reproduction, the endless drive to produce at greater and greater scales pushed by the necessity of increasing profits to survive competition. Profit can be the only determinant of production in capitalism, and the expansion of profit and therefore production is central to the survival of every unit within a capitalist system, which inherently drives towards a monopolistic state. Exchange value is at the heart of profit, and the only value that is valuable in capitalism. The contrast is with use value, what humans need and desire, but may not make a profit out of. Life and a liveable environment in themselves have no value within capitalism. This results not only in the valorisation of capital over nature, but also within environmentalism, in times of an ecological crisis, a constant effort to valorise nature through commodification. Popular constructions of the ecological crisis/climate change have three dominant characteristics: (1) 'humanity' is identified as a singular actor, (2) 'society' is identified as the sum of its parts and (3) a popular image of capitalism – There Is No Alternative (TINA) – prevails.

Mainstream environmentalism or 'bourgeois environmentalism' relies centrally on the image of a singular homogenous 'humanity' that is responsible for and has contributed to the environmental crisis we face today. While the crisis we face will, in the last instance, remain a collective one challenging our survival as an entire species, the transformation of Earth by humans is closely mediated by the organisation of society and organisation of production in particular. This has also been pointed out by critiques of the concept of the Anthropocene as mentioned earlier, who argue for the use of Capitalocene rather than Anthropocene precisely to underline this problematic framing. Popular environmentalism often remains oblivious and unwilling to address the roots of the problem, instead highlighting the symptoms, which in themselves need concern and alleviation, but are limited in providing merely symptomatic relief. Population growth, consumption, individual morality, modernity (science/technology) and imperfect markets are the most popularly identified problems.

The centralisation of population growth or consumption as the cause of the ecological crisis we face emerges from an understanding of humans as a singular undifferentiated whole rather than as a stratified society. It is ahistorical in ignoring the specific historical trajectories of colonialism and imperialism that have influenced trajectories of population growth and consumption in specific geographical regions and populations. The UN projects that world population will increase 41 per cent by 2050, to 8.9 billion people, with nearly all of this growth in developing countries. The 12 per cent of the world's population that lives in North America and Western Europe accounts for 60 per cent of private consumption spending, while the one-third living in South Asia and sub-Saharan Africa accounts for only 3.2 per cent (Worldwatch 2016). Nation-states meeting frequently to deal with climate change remain embroiled in disagreements, but generate and sustain a discourse that pits social justice and economic growth against a safe planet. Following path-breaking ideas on development as an anti-politics machine (Ferguson 1990) and a powerful discourse (Escobar 2011), it can be argued that the failure of global climate negotiations is in fact a success, measured in terms of creating a public discourse that presents a dichotomy of development versus environment, where dealing with the impending ecological crisis appears inimical to better standards of living. It frames the problem as a competition amongst nation-states, as essential differential interests of the rich and the poor states/nations/regions, all the while carefully avoiding the structures and processes that lead to differential interests. Capitalism remains unquestioned as a given, and failure seems to stem from individual personalities or abstract regional obstinacies.

Often as a direct critique of this process, moral values are raised as the terrain of struggle where the safety of humanity lies. Apart from its idealistic tendencies where the material basis of values is ignored, this perspective often tends to individualise the problem. The anthropocentric/ecocentric dichotomy emphasises different moral values that are significant but throw little light on the material bases of these values and therefore their social consequences.

Imperfect markets are seen as the bane of capitalism that results in the ecological crisis. It is argued that nature has been treated as an externality, with the market unable or unwilling to appropriately price the goods and services provided by nature. This lacuna of the market is seen as an aberration that can be fixed either through state intervention (regulatory policies) or through the market (green capitalism). While the term 'green capitalism' is of relatively recent vintage, the primary means of dealing with the ecological crisis in mainstream society for the longest time has been to rely on market mechanisms. Market environmentalism has been promoted since the early 1990s and at its core this ideology holds that the ecological crisis is because nature has not been sufficiently valued in the market. Pricing of nature's services, assignation of property rights and expansion of commodity markets into the realm of nature are the capitalist solution to the ecological crisis. Reintroducing the much-critiqued tragedy of commons (see Appell 1993; Angus 2008 for critique of Hardin 1968), market environmentalism focuses on privatising common natural resources and commodifying nature.

The incompatibility of ecosystem understanding and single value pricing pointed out by several scholars (Lele et al. 2013) remains ignored in this approach. The reductionist approach to the human-nature relationship is also worth noting here. Market environmentalism attempts to address the ecological crisis by commodifying nature. It promotes the pricing of nature's services, assigning of property rights and expansion of commodity markets to include nature's services. These solutions assume nature to be outside of humans and are unable to take into account the dialectical relationship of humans in nature.

Since exchange value is the only value that can be recognised in a capitalist economy, a variety of experiments have emerged that attempt to create exchange value for ecology. Payment for ecosystem services is one such flagship idea that has emerged to deal with the ecological crisis. Forests, rivers and clean air are provided with notional exchange values and a 'market' is created for ecological goods and services. The commodity fetishism that results from this framework that simplifies the complexity of natural ecosystems, prioritises a single exchange value and masks social relations embedded in the process of 'producing' and 'selling' ecosystem services has been highlighted by critics such as Kosoy and Corbera (2010).

Types of environmentalism: focusing on conceptual basis

As Moore (2015: 169) succinctly says, "conceptualisations of a problem and efforts to resolve that problem are always tightly connected. So, too, are the ways we think about the origins of a problem and how we think through possible solutions". It is therefore useful to distinguish environmentalism on the basis of its understanding of and approach to the crisis, rather than only on the location of its participants in economic strata. This becomes even more essential to avoid the trap of populism and essentialisation of poor as an inherently sustainable category. In practice environmental justice movements include a wide variety of perspectives and practices. While struggles over resources are significant from the point of justice, it is also important to identify and enhance their ability to resolve the ecological crisis. From this perspective, three distinct tendencies within environmentalism can be distinguished: (1) ecocentric, (2) anthropocentric and (3) metabolic. These philosophical standpoints are tendencies within social movements and are not represented here as water-tight compartments for slotting social movements. Social movements are always complex and unstable social phenomena with temporal, spatial and social variations. They may often emerge with multiple and contradictory philosophical and social locations. The classic formulation of the Narmada Bachao Andolan as an adivasi movement, or the Chipko movement as feminist have been shown to be both limited and limiting. As Baviskar (1998) has shown in the case of the Narmada Bachao Andolan (Save Narmada Movement), multiple social forces with different and sometimes conflicting perspectives, objectives and practices come together at specific times to form social movements. The categories or types of environmentalism discussed next are a heuristic device to understand specific tendencies within such movements. A blurring of boundaries between these categories at the level of empirical detail is therefore expected, and examples can only refer to certain aspects of these often complex social movements.

Ecocentrism as a philosophy emerged in the immediate aftermath of industrial capitalism responding to its monumental transformation of ecology as we knew it but also from the alienation that industrialisation engendered. It emerged as Leopold's *Land Ethic* (1949), where human exceptionalism was critiqued. It emerged as Rachel Carson's *Silent Spring* (1962), which emphasised the interdependence between human beings and nature. It is evident in Gandhi's warning that nature has enough to satisfy everyone's needs but has not enough to satisfy man's greed. Those who commit to an ecocentric philosophy hold and advocate that nature has intrinsic value, distinct from its instrumental value to the human species. A recent statement of commitment to ecocentrism avers, "the ecosphere, including the life it contains, is an *inherent good*, irrespective of whether humans are the ones valuing it" (Washington et al. 2017). Apart from claiming ethical superiority, advocates of ecocentrism also analyze the ecological crisis we face today as deriving from humanity's relentless drive towards domination of nature that results in overconsumption and overexploitation. It is the philosophical foundation of movements such as deep ecology and ecofeminism, and played a central role in the development of the field of environmental ethics.

At its extreme, it not only rejects human interest, but also rejects the very value of human existence since it is detrimental to nature. Extreme conservationist positions that argue for conservation at all costs can be seen emerging from such a philosophical position. One instance from India is the justification of some environmentalists of violent displacement of people from their homes in order to create protected areas and habitats for megafauna. Many national parks, tiger reserves and sanctuaries in India have a history of direct and indirect violence against marginalised citizens, justified or ignored in the name of nature conservation (Rangarajan 2006; Rangarajan and Sivaramakrishnan 2012). However, ecocentric philosophy is also evident in more popular versions that critique narrow ideas of valuation of nature that only recognise economic value. While many indigenous philosophies have often been classified as ecocentric since they may also recognise values beyond the economic or utilitarian, many are in fact metabolic in perspective rather than ecocentric. The distinction made here between the ecocentric and metabolic perspective is not the mere recognition of non-anthropocentric values, but the sharp dichotomy created between human interest and interests of 'nature'. The former accepts or condones harm to some sections of humanity as sad but unavoidable collateral damage while the latter sees such harm as unacceptable because there is no distinct separation of interests between human and non-human.

Anthropocentrism in its initial pre-environmental phase was seen as the source of the problem of environmental degradation. The hubris of capitalism or modernity, often undistinguished from each other, was seen as emerging from a philosophical anthropocentrism, where the only ecology that mattered was one that was of use to humans. Some of this anthropocentrism is evident in movements that argue for environmentalist action because it affects the input costs for capitalism, or the quality of life of human beings. It is worth recognising that much of inter-state negotiations at climate conferences are centred around costs and benefits to human living standards and economy.

There are also less exceptionalist versions of anthropocentrism that focus on interdependence between human beings and nature and are not substantially different from the moderate versions of ecocentrism. In this tradition, humans emerge as significant without necessarily holding them to be inherently exceptional. Many versions of the environmentalism of the poor can be seen treading this path, where livelihood rights are seen as equal to or superseding any intrinsic rights of ecological protection.

The Chipko movement can be seen as an example of a movement within the anthropocentric tradition that is popularly viewed as ecocentric through discursive constructions of activists and academia. Presently it is widely recognised that the Chipko movement in Uttarakhand where villagers, particularly women, hugged trees to protect them from felling by contractors was emergent from a long history of rights struggles of peasants (Guha 1990); that villagers were not demanding the blanket green felling ban that was subsequently imposed by the state after the movement gained global popularity as a grassroots environmental movement; and that villagers were demanding control over these forests which then rested with the state (Rangan 2000). A struggle over who controls the right to use the forest can be seen as an anthropocentric demand in particular contexts. The fact that this was an anthropocentric demand does not make it necessarily anti-ecological or any less significant for the environmental movement. However, the misrecognition of the movement as eco-centric influenced the trajectory of the movement and the outcomes, affecting both issues of justice and forest protection.

Theoretically, the metabolic perspective has a history in Marx's conception of the human-nature relationship. Foster (2000) in particular has brought back this understanding in social theory and several scholars in recent decades have developed this understanding in the context of our current recognition of the ecological crisis. For a metabolic understanding of the human-nature relationship, the basis of the ecological crisis lies in the ways that historically constituted social systems transform the relationship between natural and social systems. It begins from a focus on labour since it is through labour that humans transform the earth. However, this is never a simple unidirectional or unchanging process. The potential of human labour is realised through social organisation, and human labour produces both use value and exchange value. While human labour always transforms society, modes of organising this labour dictate what is produced, how much and for what purposes. Exchange value lies at the core of capitalism and 'value' can be understood and measured in capitalism only as exchange value. Use value becomes irrelevant in a capitalist economy.

This core understanding of the human-nature relationship that focuses not only at the species level but also engages with nature as a historical relation is found both in Marx's ecology (Foster 2000) as well as in some indigenous understandings of a social-historical relationship between human society and natural systems. The emphasis on labour in many indigenous ideologies or worldviews is often erased in their appropriation as romanticised exceptional epistemologies that appear ecocentric. This difference can be understood by drawing from feminist debates that have emphasised the significance of situated knowledge (Haraway 1991) as a counter to essentialising ecofeminism (Shiva 1989; Mies and Shiva 2010). Perceptions of the environment are implicated in practices of labour, not only at the individual level but also at the level of social organisation.

Savyasaachi's study among the Hill Kharias in Simlipal provides one such example. He writes, "The forest dwellers recognize a life-force which flows through the forest, through its food chains and life-cycles, from which all the elements of the forest, including man, draw nourishment" (Savyasaachi 2007: 476). Unlike the centralised management of Simlipal as a tiger reserve, which seeks unsuccessfully to create discrete geographies that separate metabolic flows (ecocentric, as described earlier), Kharia perceive the interconnected inherent in the ecosystem and see the forest as a workplace.

Similar to Liebig's description of the impact of capitalist agriculture on soil fertility, in Simlipal, circumstances created by Project Tiger interfere with the life-cycle of plants, disrupt the food chain and prepare the ground for deforestation (ibid.). Savyasaachi describes how the detritus food chain, which orders the process of self-regeneration, is disturbed. The collection of sal seeds reduces the food material for herbivores and saprophytes, which in turn leads to the decline of population of rodents, squirrels, porcupine, hare and rabbits. This in turn affects the population of reptiles. The collection of sal leaves affects the thickness of the humus layer on the forest floor, which in turn affects growth of mushrooms and toadstool. With a decrease in mushrooms, people in Simlipal look for wild animals for their protein nutrition, which results in their decline. This case study presents one example of an indigenous metabolic perspective.

Towards a conclusion: addressing the metabolic rift

Classification is always about power, and it conceals as much as it reveals. The classification of environmentalism is equally implicated in this project of classification. The Environmentalism of the Poor was a particularly powerful project of classification that drew attention to inequality as a central concern and provided the framework for the environmental justice movement. At its radical best it drew attention to an imperial history that is implicated in our current ecological crisis. It established the link between justice and sustainability. However, as with many radical concepts, its mainstreaming also resulted in its declawing. While movements that retain their radical edge continue to struggle for environmental justice, the widespread use of the concept also conceals and underplays the implication of capitalism in the ecological crisis.

The definition of ecological problems significantly prefigures their solution. In understanding the Anthropocene, in responding to climate change, in the ability of social movements to respond to the ecological crisis, how we understand the problem we want to address is significant. This chapter has established how the metabolic rift is conceptually significant to understand environmental movements. The metabolic rift is an important concept not only as a critique of historical capitalism, but as a framing of our current ecological crisis. Our hope of success in efforts to address the breaching of planetary boundaries and our survival as a species lies in addressing this fundamental rift.

Note

1 This chapter was originally published as Sudha Vasan, 'Ecological Crisis and the Logic of Capital', *Sociological Bulletin*, Vol. 67, Issue 3, pp. 275–289. Copyright 2018 © Indian Sociological Society. All rights reserved. Reproduced with the permission of the copyright holders and the publishers, SAGE Publications India Pvt. Ltd, New Delhi.

References

Adger, N.W. (Ed). 2006. *Fairness in Adapting to Climate Change*. Cambridge, MA: MIT Press.

Angus, Ian. 2008. *The Myth of Tragedy of the Commons*. Available at http://climateandcapitalism.com/2008/08/25/debunking-the-tragedy-of-the-commons/. Last accessed on 20/05/2017.

Appell, G.N. 1993. *Hardin's Myth of the Commons: The Tragedy of Conceptual Confusions*. Working Paper 8. Phillips, ME: Social Transformation and Adaptation Research Institute.

Baviskar, Amita. 1998. *In the Belly of the River: Tribal Conflicts over Development in the Narmada Valley*. New Delhi: Oxford University Press.

Baviskar, Amita. 2003. Between Violence and Desire: Space, Power, and Identity in the Making of Metropolitan Delhi. *International Social Science Journal, 55*: 89–98.

Burkett, Paul. 1999. *Marx and Nature: A Red and Green Perspective*. New York: Palgrave.

Byravan, Sujatha and Sudhir Chella Rajan. 2010. The Ethical Implications of Sea-Level Rise Due to Climate Change. *Ethics & International Affairs*, 24(3): 239–260.

Carson, R. 1962. *Silent Spring*. New York: Mariner Books, Houghton Mifflin Company.

Clark, B. and J.B. Foster. 2009. Ecological Imperialism and the Global Metabolic Rift: Unequal Exchange and the Guano/Nitrates Trade. *International Journal of Comparative Sociology*, 50(3–4): 311–334.

Escobar, A. 2011. (Revised edition, first edition 1995) *Encountering Development: The Making and Unmaking of the Third World*. Princeton, NJ: Princeton University Press.

Ferguson, J. 1990. *The Anti-Politics Machine: "Development", Depoliticisation and Bureaucratic Power in Lesotho*. Cambridge: Cambridge University Press.

Foster, J.B. 1999. Marx's Theory of Metabolic Rift: Classical Foundations for Environmental Sociology. *American Journal of Sociology*, 105(2): 366–405.

Foster, J.B. 2000. *Marx's Ecology: Materialism in Nature*. New York: Monthly Review Press.

Foster, J.N., B. Clark and R. York. 2011. *The Ecological Rift: Capitalism's War on Nature*. New York: Monthly Review Press.

Guha, R. 1988. Ideological Trends in Indian Environmentalism. *Economic and Political Weekly*, 23(49): 2578–2581.

Guha, R. 1990. *The Unquiet Woods: Ecological Change and Peasant Resistance in the Himalaya*. Berkeley: University of California Press.

Guha, R. 2000. *Environmentalism: A Global History*. New York: Longman and New Delhi: Oxford University Press.

Guha, R. and J. Martinez-Alier 1997. *Varieties of Environmentalism: Essays North and South*. London: Earthscan and New Delhi: Oxford University Press.

Haraway, Donna. 1991. *Simians, Cyborgs and Women: The Reinvention of Nature*. New York: Routledge.

Hardin, G. 1968. The Tragedy of the Commons. *Science*, 162(3859): 1243–1248.

Harding, Sandra. 1986. *The Science Question in Feminism*. Milton Keynes: Open University Press.

Hornborg, Alf. 2003. "The Unequal Exchange of Time and Space: Toward a Non-Normative Ecological Theory of Exploitation." *Journal of Ecological Anthropology* 7, (1): 4–10.

Intergovernmental Panel on Climate Change (IPCC). 2007. *Climate Change 2007 – Impacts, Adaptation and Vulnerability: Contribution of Working Group II to the Fourth Assessment Report of the IPCC.* New York: Cambridge University Press.

Jaggar, Alison. 1983. *Feminist Politics and Human Nature.* Totowa, New Jersey: Rowman and Allenheld.

Kosoy, N. and E. Corbera. 2010. Payments for Ecosystem Services as Commodity Fetishism. *Ecological Economics*, 69(6): 1228–1236.

Lele, S., O. Springate-Baginski, R. Lakerveld, D. Deb and P. Dash. 2013. Ecosystem Services: Origins, Contributions, Pitfalls, and Alternatives. *Conservation and Society*, 11(4): 343–358.

Leopold, A. 1949, 1968. *A Sand County Almanac and Sketches Here and There.* Oxford: Oxford University Press.

Martinez-Alier, J. 2003. *The Environmentalism of the Poor: A Study of Ecological Conflicts and Valuation.* Cheltenham, UK and Northampton, MA: Edward Elgar.

Mies, M. and V. Shiva. 2010. *Ecofeminism.* New Delhi: Rawat Publications.

Moore, Jason W. 2015. *Capitalism in the Web of Life: Ecology and the Accumulation of Capital.* New York: Verso.

Rangan, H. 2000. *Of Myths and Movements: Rewriting Chipko in Himalayan History.* London and New York: Verso.

Rangarajan, M. 2006. *India's Wildlife History: An Introduction.* New Delhi: Permanent Black in Association with the Ranthambhore Foundation.

Rangarajan, M. and K. Sivaramakrishnan (Eds.). 2012. *India's Environmental History.* Volume I: From Earliest times to the Colonial Era. Volume II: Colonialism, Modernity and the Nation. New Delhi: Permanent Black.

Rockstrom, J., Steffen, W., Noone, K., et al. 2009. Planetary Boundaries: Exploring the Safe Operating Space for Humanity. *Ecology and Society,* 14(2): 32.

Savyasaachi. 1994. The Tiger and the Honeybee. *Seminar*, 423: 30–35. Reprinted in Rangarajan, M. (Ed.). 2007. *Environmental Issues in India: A Reader.* New Delhi: Pearson Education India.

Schoenfeld, S., 2005. Types of environmental narratives and their utility for understanding Israeli and Palestinian environmentalism. In *Palestinian and Israeli Environmental Narratives: Proceedings of a Conference Held in Association with the Middle East Environmental Futures Project* (pp. 93-114). Toronto, Ontario: Center for International and Security Studies, York University.

Shiva, V. 1989. *Staying Alive: Women, Ecology and Development.* New Delhi: Kali for Women and New Jersey: Zed Books.

Shue, H. 1999. Global Environment and International Inequality. *International Affairs*, 75(3): 531–545.

Steffan, W., P.J. Crutzen and J.R. McNeil. 2007 The Anthropocene: Are Humans Now Overwhelming the Great Forces of Nature. *Ambio*, 36(8): 614–621.

United Nations Development Programme (UNDP). 2011. *Human Development Report 2011. Sustainability and Equity: A Better Future for All.* New York: UNDP.

Washington, H., B. Taylor, H. Kopnina, P. Cryer and J.J. Piccolo. 2017. *Statement of Commitment to Ecocentrism.* Available at www.ecologicalcitizen.net/statement-of-ecocentrism.php. Last accessed 05/04/2017.

Wolstenholme, Ruth. 2009. *Differential Social Impacts of Climate Change in the UK*. Research Report. Available at www.knowledgescotland.org/briefings.php?id=95. Last accessed on 27/02/2012.

World Commission on Environment and Development (WCED). 1987. *Our Common Future*. Oxford: Oxford University Press, p. 27.

Worldwatch. 2016. *The State of Consumption Today*. Available at www.worldwatch.org/node/810#3. Last accessed on 05/04/2017.

3

NATURE, NATION, SCIENCE AND GENDER

Gita Chadha[1]

Introduction

I loved Science, Stars, Nature, but then I loved people without knowing that people have long since divorced from nature. Our feelings are second handed. Our love is constructed. Our beliefs colored. Our originality valid through artificial art. It has become truly difficult to love without getting hurt.

Rohith Vemula, Suicide Note, 2016[2]

*

The genesis of this chapter lies in the keynote address that was scheduled and presented by Mahesh Rangarajan at the seminar Reframing the Environment: Resources, Risk & Resistance in Neoliberal India, organised at Mumbai University in January 2016. His book (Rangarajan 2015) titled *Nature and Nation: Essays in Environmental History* set me thinking about the complex network of issues pertaining to our ideas of an environmental history. I realised that environmental issues that plot the dots between nation and nature need to be informed by two other dots: science and gender. Insights from my own engagements with feminism and science clearly indicated the fact that environmental and feminist movements intersect to critique not just the nation state but also science as a *raison d'etre* of the nation state (see Introduction of Nandy 1989). More importantly, women's role as knowledge producers and as keepers of nature were important themes that have constituted the environmental movements in India. Rangarajan states that "there was no simple 'Western' model of nature transplanted into India (as indeed of nation)" (p. 32) and seems to suggest that the control of the modern nation state over nature was a hybridised gesture aimed at development at the cost of peace with nature. He states, "peace with nature, howsoever defined, has to be concomitant with and underpinned by peace among people" (p. 36). I was intrigued by the fact that neither gender nor science feature, even cursorily, in Rangarajan's otherwise excellent and well-intentioned account of environmental history. The possibility of bringing science and gender into the discourse became an exciting proposition for me to explore.

One of the defining arbiters in the social and material relationship between nature and people has been modern Western science. Critically tied to the project of development and progress, modern Western science has been paradigmatic in envisioning nation states. Interestingly, nationalist discourses in India have imagined and represented the nation as the mother, feminised in and through multiple discourses. With the development of science studies, especially after the 1960s, science has been critically examined and re-imagined from several locations. Feminist and post-colonial critiques have been important contributors to the development of science criticism, just as pacifist and environmental critiques have been. Simultaneously, feminists have also critiqued the feminification of the nation (see, for example, Krishnaraj 2010). A critical examination of these intersections provides insights for feminist studies of nation and nationalism, for feminist studies of science and for feminist environmental studies. In this chapter, I aim to explore these intersections in the Indian context.

Gender symbolism in science and nationalism

Feminists have critically examined the epistemological distinctions between mind and matter, rationality and intuition, objectivity and relatedness, neutrality and value-imbuedness. Feminists argue that these dichotomies are the basis of modern Western science which evolved out of a conceptual structuring of the world that incorporated particular and historically specific ideologies of gender too. These dichotomies associated with science are also associated with the making of gender binaries. It is often argued by the post-modernists that these dichotomies are constructs – often of language – and have no ontological 'truth' value attached to them, thus bringing into question the very basis of modern Western science. On the other side, the defenders of the 'neutrality' of science say that these dichotomies are 'just' language – some sort of heuristic tools – used to describe a pre-existing ontological universe. Addressing this tricky question of the relationship between language and truth, Genevieve Lloyd states that "the male-ness of the man of reason is no superficial linguistic bias" (Lloyd 1996: 41) but has deeper roots within the Western philosophical tradition. "Rational knowledge", she says, "has been construed as a transcending, transformation or control of natural forces; and the feminine has been associated with what rational knowledge transcends, dominates or simply leaves behind" (Lloyd 1996: 41). Therefore, what was left behind in the project of modernity was always symbolised as feminine and female. Apart from this epistemological critique, the feminist critique of science also makes the radical move of arguing that the content and theories of natural science are gendered. Though this is not an easy task, there seems to be a greater possibility of demonstrating this in sciences like biology and primatology than in physics and mathematics.

More generally, discussing the tacit gender symbolism present in the ontological assumptions made by science, Sandra Harding writes that "gender symbolisations generally occur in the margins, in the asides, of texts – in those places where speakers reveal the assumptions they think they do not need to defend, beliefs they expect to share with their audiences" (Harding 1996: 112). This gender symbolism prevails through (a) how nature is constructed as the ontological object, (b) how the mode and method of inquiry, i.e. the scientific method, is constructed as 'objective' and (c) how the human 'subject' – the inquirer, the scientist – is constructed as the masculinised conqueror/genius/knower, in this case of nature and of nation. This gender symbolism, according to feminist science critics, contains an order and ideology, a thought and practice, that becomes "a resource for the advancement of science" and vice versa. It is often expressed in and as 'metaphor'. Therefore, investigating metaphors has been an important method in uncovering gender symbolism.[3]

A similar symbolism – not as tacit as it is in science – prevails in the construction of South Asian nationalisms. I take India as a specific case. The feminification of nation as mother and the relational masculinisation of the male citizen, the patriot, as the son of the soil who has to safeguard the interest of the motherland and of mother nation, often even sacrifice his life for her, are commonly found in the popular culture and in the common-sense universe of modern and contemporary India. Evidenced in early nationalist literature and iconography of the nation, the mothering of the nation is most entrenched in the widely popular rallying song of Indian Nationalism 'Vande Mataram', played even today. In fact, though popular across governmental regimes of varied ideologies, the song is seeing a revival in the present millennium as a right-wing cultural nationalism soars. Interestingly, popular cinema in dominant regions and languages reproduces the idea of a 'Mother India', the name of a classic 1950s Hindi film. With a specific focus on Bengal, as Jashodhra Bagchi states, "in order to lend force to nationalism, the ideology of motherhood was given an enormous importance in the cultural life of Bengal" (Bagchi 2010: 159). Since the mother in a patriarchal culture needs to be legitimised as the mother of sons, Bagchi argues that largely "the hagiography of all social reformers contains eulogies of their mothers. Mothers were justified by the greatness of their sons" (Bagchi 2010: 161). Pointing out that this kind of gender symbolism around the glorified mother was actually a kind of 'compensatory history' that was being constructed by the Indian nationalists and social reformers of the later period, particularly in Bengal, Bagchi states that "The glorification of motherhood in colonial Bengal was merely in the domain of ideology. Such an ideology", she argues "was based on a philosophy of deprivation for women in the world of practice" (Bagchi 2010: 159). It is within this gender symbolism in the construction of nature and nation, tacit and direct, that I place the question of science through an exploration of an important scientific figure of the time, the eminent biologist J. C. Bose. Ashish Nandy, in his classic *Alternative Sciences* proposes that reading Bose's life and works might reveal to us an interesting hybrid practice of how science was transplanted into the mind and practice of the scientific intellectual of the time. Interestingly, Madhumita Mazumdar points out that the scientific intellectual was constructed both as Acharya, the learned one, and rishi, the ascetic one (Mazumdar 2015). The aim of the chapter here is to suggest that while the 'alternative' suggested by Bose helps feminist reformulations of nature, it still does not free us form gender associations.

Nature, gender and the birth of modern Western science

In a section titled "Should the History and Philosophy of Science Be X-Rated?" in her classic *The Science Question in Feminism* (Harding 1996), Harding discusses five changes, first articulated by Carolyn Merchant (1980), in the relationship between nature and people, in thought and experience, that mark the birth of modern Western science. These changes that characterised thought and social experience in Europe in between the 15th and the 17th centuries contributed to the gendering of the subsequent scientific worldview that emerged.

First, Merchant suggests that the shift to a heliocentric view of the galaxy from a geocentric one had significant consequences on the construction of a gender ideology. She argues that this shift led to the displacement of the organic, renaissance, vitalistic representation of nature as earth. While in this worldview, the earth was feminified either as nurturing mother of sons or as a wild, uncontrollable female nature, in the new Copernican worldview, it was reduced to a small irrelevant planet, amongst many others, that circles around the sun. The centrality of the 'earth' disappears.

The second change is the return of the Aristotlean view of the passive as feminine and active as masculine[4] in 16th-century science. This came along with the scientific construction of the difference between inanimate and animate. This change led to the making of a passive inanimate earth that is fertile only with a 'marriage' of the earth and the sun, best represented in the Copernican metaphor of "the earth conceives by the sun and becomes pregnant with annual offspring" (Harding 1996: 114). Both these shifts established the power of science over nature, where nature became an object to be controlled through experimentation and harnessed through intervention.

The older Aristotlean-Ptolemiac conception of the universe separated the earth and the heavens – the terrestrial and the celestial. The celestial sphere was not subject to decay and dissolution and was unchanging and eternal. With the movement to the Copernican worldview, the distinction between the terrestrial and the celestial worlds was completely obliterated and the heavens were as susceptible to change as the earth. The possibility of change in the heavens which were so far eternal and immutable led to a shift in the conception of nature as unruly and chaotic, which had to be controlled and put to use by the mind of man. This led to a major conceptual shift in the man-nature relationship, the third change that Merchant lists.

The fourth change that Merchant draws our attention to is traced to the changing social relations of 15th- and 16th-century Europe. As though mirroring the heavens, the social structure of Europe in the 15th and 16th centuries was also seeing major changes and upheavals due to the breakdown of the feudal society. In tandem with new movements like Protestantism, this led, in particular, to the increased visibility of women in public life. On the one hand was a firmly established equation between nature, disorder and chaos with the feminine and, on the other, there was the Renaissance equation of the animate and the inanimate, of the physical and the social. It was only one simple step of deduction from here to conclude that all disorder, natural or social, is to be associated with women. From this rich brew of beliefs and conception that were stirred together in the 16th century cauldron emerged the discourse around witchcraft – a potent way to control and subjugate women.

Witchcraft, the political and legal discourses around it and the attacks and subjugation of women under the guise of dealing with witchcraft was, according to Merchant, the root of the fifth and the last change leading to the gendering of the scientific method. In fact, it is not a coincidence that early writings around the method of science happened concomitantly with the legislation against witchcraft under the rule of James I, the first king of Great Britain and a strong proponent of anti-witchcraft legislation. King James' trusted lieutenant, advisor and confidante was Sir Francis Bacon who is also credited with the first systematic account of the method of science. Bacon's views on empiricism have become the holy grail of modern science. Bacon's writing on science generously dips into the anti-witchcraft thinking of the time to draw upon a fount of gendered metaphors. In a striking passage addressed to the king, Bacon writes:

> For you have but to follow and as it were hound nature in her wanderings, and you will be able when you like to lead and drive her afterward to the same place again. . . . Neither ought a man to make scruple of entering and penetrating into those holes and comers, when the inquisition of truth is his whole object- as your majesty has shown in your own example.
>
> (Harding 1996: 116)

Moreover, due to the large and dominant presence of men in science and in the audience for science – in the public sphere and domain – Harding states that "the best scientific activity and philosophical thinking about science are . . . modelled on men's most misogynous relationships to women-rape, torture, choosing 'mistresses', thinking of mature women as good for nothing but mothering" (Harding 1996: 112).

There are significant things that happen as this modern science is transplanted across the world through colonialism. For the context of the present chapter, I extend Merchant's feminist critiques to problematise another axis of the feminification of nature – and as a consequence the masculinisation of culture (and science) through metaphors. In the Indian context, I would like to critically examine the construction of the discourse around mother, mother goddess and science. For this, I am picking out a defining moment in the making of the Indian nation and Indian modernity – the colonial national moment. In this formation, I submit that nature and nation were simultaneously represented and constructed as a deified, yet disempowered, feminine. Nature and nation were co-constructed so to speak, within a gender symbolism that encoded layers of patriarchy. Most importantly for this chapter, I suggest that this process was fundamentally locked with the transplanting of science in India.

J. C. Bose: an alternative approach?

In order to examine the triad of science, nature and nation in the Indian context, I choose the life and work of J. C. Bose as the focal point of my discussion. J. C. Bose becomes almost a natural choice for my study because he inhabits very centrally the intersection of the science-nature-nation triad. Some of his work in science has been recognised as very important but, at the same time, the part of his work that ran counter to mainstream science ideas are now largely discounted. His efforts that went against the mainstream approach stemmed out of a new conception of science itself that Bose had come up with, synthesising the modern science of the West with the wisdom of ancient India. In this re-imagination of science, Bose was negotiating the tensions that were apparent between the conception of nature in modern science and in ancient Indian traditions. It is for this reason that Bose becomes the focus of my attention.

In his work on Bose (Nandy 1980), Ashis Nandy has suggested that Bose's work, given the colonial context it emerged in and the tradition that it emerged from, represents an alternative to the modern, European conception of science. I want to pursue this further and try to examine Bose's conception of a pan-vitalist science and see if it carries within itself a feminist consciousness or if it is at all possible for feminists to insert a feminist reading, thereby claiming it, in the sense that Harding suggests. Specifically, I would like to view Bose's conception of nature and nation through the lens of gender to see if it might have consonance with the re-visualisation of science that feminists suggest as 'alternative' to what they critique.

Vitalism

In a monograph, published in 1902, called the *Response in the Living and the Non-living*, Bose used the ideas inspired by Waller's work on electrophysiology to use electrical response as the indicator of life (Dasgupta 1999) – or even, as he called it in a letter to Tagore, as the 'ultimate sign of life'. Waller himself had studied electrical response in various living systems but had been more guarded in using his work on the response of living systems to electrical impulses to make it a sufficient condition for life. This is precisely what Bose did and he went further to assert a continuum between living and non-living systems. The Boseian thesis, as it has been called, is that there is no demarcation between the living and the non-living but, in fact, they exist on a continuum. Bose, it can be claimed, occupied an altogether new position in the debate between mechanism – a reductionist position seeking to reduce all of life to physics – and vitalism, which claimed that living systems are possessed with a special quality that cannot be reduced to non-living material. Bose's own position was a form of pan-vitalism where living and non-living matter were on a continuum but the non-living matter was possessed with a vitality. So even though he was not essentialising the difference between the living and the non-living like the vitalists, he was antipodally positioned with the materialists because he considered non-living matter as also being imbued with the quality of vitality. In a significant statement of this philosophical position in an address to the Royal Institute, Bose says:

> I have shown you this evening the autographic records of the stress and strain both the living and the non-living. How similar are the two sets of writing, so similar indeed that you cannot tell them one from the other? They show you the waxing and the waning pulsations of life – and climax due to stimulants, the gradual decline in fatigue, the rapid setting in of death rigour from the toxic effect of the poison. It was when I came on this mute witness of life and saw an all-pervading unity that binds together all things – it was then that for the first time I understood the message proclaimed on the banks of the Ganges thirty years ago.
>
> (Quoted in Nandy 1980: 62)

The connection that Bose makes in this statement is striking because it traces in one quick movement the origin of the idea of the unity of the living and the non-living to some unnamed Indian roots.

Delving into Bose's understanding of this continuum, it becomes inevitable to find the connections with the philosophy of Sankhya, which he was deeply influenced by.[5] The fundamental realities that constitute the universe are *purusa* (consciousness) and *prakriti* (nature/matter), the male and female principles, respectively. *Prakriti* in this system is the active principle responsible for the manifest universe which is governed by strict laws of causality. The description of *prakriti* in Sankhya literature is often as an unintelligent principle which is ruthless and uncaring because it is driven by strict causal principles. This is starkly in contrast to the benevolent and nurturing feminine principle invoked in systems like *sakta* or *tantra*, which was also something that had a hold on Bose's consciousness. In his essay *Abyakto*, Bose talks of his understanding of *prakriti* as 'motherliness in the steel-frame of inexorable orderliness'. The mother metaphor gets clearly articulated in Bose's perception of nature as feminine, nurturing and caring.

However, as with many Indian intellectuals in the 19th and early 20th centuries, Bose's understanding of Indian philosophy was a weaving together of many strands, making for a consequent eclecticism. This is evident in the fact that Bose was also engaged with Vedantic non-dualism, a school of thought positioned as contrary to the dualism of Sankhya. One could trace this strand of Bose's make-up to his father who had converted to the monotheism of the Brahmo Samaj though he belonged to a family of East Bengal Brahmins who were followers of Sakta religious practices. The mother-goddess of these practices presented a somewhat different feminine principle than that of Sankhya; she was, at once, an embodiment of terrifying power and of loving care. It is this form of the feminine divine that the young Bose might have internalised through his exposure to the myths and tales. Ashis Nandy even attempts (Nandy 1980) to find the roots of Bose's understanding of nature as *prakriti*, on the one hand, and as the *sakta* goddess, on the other, deep in Bose's subconscious and his traumatically ambivalent relationship with his mother. Nandy further suggests that of Bose's practice of science were both personalised and revelatory, something that the Western practice of science had left behind for the more instrumentalist and impersonalised practice, over determined by the hypothetico-deductive method (Nandy 1980: 85).

Nationalism

I would now like to move to establishing another aspect of my thesis: Bose's conception of nature was intertwined with his ideas of nation/nationalism and that his science directly stemmed out of his philosophical position outlined earlier. Nandy, in fact, argues that Bose's conception of science and its method was radically different from the positivistic ideas of science that the West was steeped in. He argues that for Bose science was not a matter of discovery but one of revelation. As Nandy says, "he conceived of science as a means of experimentally demonstrating the truth of self-evident, axiomatic, general laws of nature enunciated in the sacred texts of India" (Nandy 1980: 84). The intellectual climate of the late 19th and early 20th century in India and, particularly in Bengal, was such that many Indian intellectuals sought a sense of national pride in the wisdom of ancient India. Bose was one of the earliest, however, who linked this wisdom to and equated it with the modern science of the West. Through this equation, scientific knowledge was not an empirical enterprise understood in the usual sense but a revelatory experience. In the cultural context that Bose was placed, the act of revelation was usually one mediated by a divine mother to the seeking devotee. The epistemological process of knowledge production in science was, in this convoluted manner, gendered and this gendering was not just in spite of the universalism that Bose advocated but it was, in fact, because of it. This gendered conception of nature, which informed Bose's science, was intertwined, as I said earlier, with the idea of nation, which as we will argue later is also gendered. The intertwining of nature and nation thus leads to a double gendering.

Feminification of the nation: gender symbolism and the visual in India

Maitrayi Krishnaraj in her introduction to the book *Motherhood in India* writes that the woman as mother is a glorified icon in literature, popular culture and the everyday world. As she says, "it is not the fact of mothering that makes women vulnerable, but their social construction, the implications for women flowing from the meaning attached to the idea of motherhood, and the terms and conditions under which it is allowed to express itself" (Krishnaraj 2010: 7). In the process of feminification, when gendered metaphors are used to describe aspects of our experience, then in that act of symbolising gender biases are also taken forward. In describing the nation as mother, for example, the entire gendered conception of motherhood is transplanted onto nation. As a result, the nation as a symbol is made vulnerable. The gendered conception of nation as mother persists through the different ways in which the nation is apprehended: as an abstract entity, or in its physical manifestation as land or soil, or through its representation in literature, culture and in cartography.

Sumathi Ramaswamy in *Visualising India's Geo-Body: Globes, Maps, Bodyscapes* states that "the national longing for cartographic form in colonial and post-colonial India . . . finds expression in the visual practices of patriotism" (Ramaswamy 2003: 151–152). She argues that maps of the national territory are compelling, peculiar and novel technologies of persuasion used by modern patriots. Distinguishing a disenchanted state cartography from the enchanted cartography of the patriot, Ramaswamy argues that modern patriotic representations of the nation are feminised in hybrid forms. She analyses images where the mother goddess images are inserted into the bodyscape of the Indian map. She writes, "by the time the Ghadar Party published the image of Bharat Mata in 1923 . . . what had begun as an occasional and imprecise practice in the 1860s of imagining India as a female entity had matured into a routinised habit" (Ramaswamy 2003:

172–173). Analysing the visual cartography of the modern patriot, Ramaswamy takes us through the representation of nation as the 'mighty shakti' of Aurobindo Ghosh and the 'enchanted somatic imaginations of India' as articulated by Nehru. Further, Ramaswamy argues that in order that this Mother India be owned, the impersonal territorial cartography was infused with the insertion of its sons:

> It is clear that, for the patriot, "the nation" could never just amount to the territory delineated by colonial geography and cartography and to the impersonal topographic elements that constituted it. These necessarily had to be supplemented by the somatic imagery of Bharat Mata and of her loyal "sons", who were wedded to the soil from immemorial generations.
>
> (Ramaswamy 2003: 177)

Jashodhara Bagchi in her work *Representing Nationalism: Ideology of Motherhood in Colonial Bengal* argues that "in order to lend force to nationalism the ideology of motherhood was given an enormous importance in the cultural life of Bengal" (Bagchi 2010: 159). She further writes that this ideology served to disempower women and was based on a philosophy of deprivation. She persuasively demonstrates that it was not only the nationalists but also the liberal social reformers who were responsible for bringing in the mother metaphor to represent the nation.[6] Bagchi argues that it was not as if the liberal social reformers, the modernists, were better than the later nationalists on the gender issue or on mother glorification. Bagchi sees the celebration of a mother through her sons as the "compensatory history" of women (Bagchi 2010: 161). Bagchi further suggests the equation between a tender mother and soil in Bengal established a gender symbolism "that encouraged the representation of the Bengali woman as an affectionate mother, ever ready to respond to the demands of her children" (Bagchi 1990). She states, "With the emergence of Bengalis as a distinctive identity in the Gangetic delta, a confirmation of the spirit of tender motherhood was found in the natural setting which Bankimchandra described as well watered and fertile (*sujalam-suphalam*)". The mother earth was a healer of sons and their humiliation. She was both kind and fiery, often needing protection.

Madhumita Mazumdar (Mazumdar, 2017, p. 185) suggests that it was the cultural-nationalist framing of science that marked its alignment with emerging discourses of gender and class in Bengal. Tagore, in a letter to Bose, writes:

> Dear friend, You ventured into the Goddess of science's beloved western shrine. And returned in triumph to garland the lowly head of your distressed motherland. Today she sends you her blessings

> through the voice of an unknown bard. And hopes that its faint reverberations reaches and touches your heart.

This identification of mother and earth/land/nation finds direct resonance in the complex worldview that Bose inherited and developed. "Can a son imagine a distinction between motherland and mother?" he asked, rhetorically, in a letter to Subhas Chandra Bose he wrote in 1937. This was a firmly established equation for Bose and was foreshadowed with all the complexity that his traumatic relationship with his mother, on the one hand, and his nurturing relationship with his wife, on the other, had left him to grapple with. His father's early demise and his mother's widowhood were to mark him indelibly. In a letter to Tagore that he wrote from London, he talked of a vision that had strengthened his love for his motherland:

> All of a sudden I saw a shadowy figure, wearing the dress of a widow; I could see only one side of the face. That very sickly, very unhappy woman's shadow said, "I have come to accept"; then, within a moment, the whole thing disappeared.

The impoverished nation was the widow, impoverished by centuries of colonial rule. But, for Bose, paradoxically, it is science that is seen as the panacea for the ills of a colonial nation – the same science that tames wild, unreasonable nature is also the path to the emancipation of the motherland. In this double-gendering of nature and nation that happens, Bose displays an ambivalence toward science, too. He attempts to not just use it as an import from the West but also to infuse it with a new consciousness that is very Indian. His researches in science, he felt, were taking him back to an understanding that was essentially Indian. The unity that he saw between living and non-living systems offered him justification for the philosophical framework which he identified as that of his motherland. The son of the soil seems more predominant here than the citizen of a nation.

Conclusion

In this concluding section, I wish to bring together the various strands of argument that have been explicated in the course of this chapter. First, I would like to evoke the concept note that framed the seminar Reframing the Environment: Resources, Risk and Resistance in Neoliberal India, to which this chapter was a contribution. I would then like to return to the discussion in the introduction with respect to the keynote address of Mahesh Rangarajan, my own departure from his position and how the analysis presented of J. C. Bose in this chapter helps shed light on these questions. I will finally discuss what directions can be explored as alternatives to our present understanding of the science, nature and nation.

The seminar Reframing the Environment: Resources, Risk and Resistance in Neoliberal India sought to explore the politics, especially the neoliberal politics, that underlies and frames much of the present-day discussion on environment, ecology and sustainable growth. The concept note of the seminar explicitly stated its aim as "to unravel the underlying power relations that are masked and hidden in the present discourse of ecological sustainability". There are many directions from which to address such an aim – I have chosen to pitch this question not within the domain of either economics or geo-politics but at the more elementary level of science and development itself. I argue that the development question in science is fundamentally tied up with the process of knowledge production in science and that the issue of growth and development brings us back to asking how science fundamentally understands nature. The issue of development also necessarily brings in the related issue of the nation, as pointed out by Mahesh Rangarajan in the keynote address to the seminar. I have demonstrated in this chapter that, at a fundamental level, both nature and nation have been feminified in scientific modernity and, therefore, it is necessary to look at these questions from a feminist perspective. Feminist theories of science have provided us with revolutionary insights into the triad of nature, nation and science and affect a reworking of all the attendant questions like that of development.

Just as Rangarajan has provided us a new window by compelling us to look at nation and nature simultaneously, Ashis Nandy in his famous book *Alternative Sciences* presents an account of the life of the scientist Jagadish Chandra Bose and argues that his conception and practice of science challenge the modern, Western understanding. I argue that, just as it was necessary to complicate the nature-nation question Rangarajan poses by bringing in gender, one also has to look at J. C. Bose's proposed alternative conception of science from a gender perspective. My reading of Bose suggests that while his science may provide an alternative as far as the colonial question is concerned, it does not provide a true alternative from a feminist perspective. If offered as an alternative to Western models of scientific practice, the panvitalist science along with the sentimental nationalism of Bose as a modern patriot seem problematic. They reproduce gender symbolisms which are deeply patriarchal. The idea of a feminine Nature, wild and devoid of reason, tamed by the Man of Reason makes the transition from being a metaphor to a metaphysic. Contrasted with this is the vitalist position which seeks to break this duality but, in doing so, sometimes brings in a new set of gendered metaphors. A feminist approach to nature would veer towards the vitalistic but after having avoided the pitfalls of these metaphors. Feminist environmental movements and feminist science studies have shown how such a vitalist position is possible. This position allows the production of a new narrative of science which focuses on issues like sustainability and care rather than productivity and exploitation and produces a less alienated account of nature itself. In this mechanistic/vitalistic duality when we now introduce a third position of Bose's pan-vitalism it appears, at first sight, that this is a position completely at variance with a mechanistic conception of the world and thereby automatically synchronised with the feminist critiques of a mechanistic conception of nature. However, and paradoxically, Bose's pan-vitalistic view reproduces the same set of gendered dualities that the mechanistic view does. This is because, as explained earlier, Bose's pan-vitalism is itself imposing patriarchal gender dualities. Whether nature is seen as the *prakriti* in Sankhya or as the mother following Sakta traditions, the male-female duality is all-pervasive. Bose's pan-vitalism therefore does not, I would submit, contain a feminist consciousness that we may draw upon.

For a more robust alternative to modern Western science, an alternative based upon non-binary ideas of the world must be enabled. For this, I would draw upon Elizabeth Fee's (Fee 1981) argument that the Cartesian dichotomy set up by modern science needs to be broken because it only serves to legitimise the hegemony of modern science over other ways and methods of knowing and practice. Fee argues that there are four dichotomies, from the Cartesian list, that a feminist approach to science must destabilise.

The first dichotomy is the one between knowledge for knowledge and knowledge for social uses. Retaining this dichotomy, according to her, allows the scientist to absolve themselves from any social responsibility. While we would say that science is not 'pure' knowledge and is always sullied and influenced by the social context of production, we also recognise its powers to construct/represent the world. In this sense, using this 'impure' nature of science, we must demand accountability and responsibility from scientists. Second, Fee argues that the division between thinking and feeling also promotes a false consciousness whereby scientists are made to believe that their emotional involvement must exist only in their personal lives and must not spread to their thinking selves, almost leading to a fractured, fragmented and schizophrenic self and practice amongst scientists. Third, Fee argues that 'subject' i.e. the knowing subject is the scientist who is active and who asks questions, whereas the object of scientific knowledge is passive, speaks only when spoken to and only as much as is required to respond to the questions. A more hermeneutic approach is what we would advocate, where we learn or re-learn to listen to nature. Lastly, Fee argues that the distinction between science and society must be destabilised to reveal the political nature of science, to see and show how science is deeply implicated in the politics of nations and vested interest.

All these steps, I think will lead us to the path of the love that someone like Rohith dreamt of, was perhaps deprived of: a path of pluralism, inclusion and compassion.

Notes

1 This chapter was originally published as Gita Chadha, 'Nature, Nation, Science and Gender', *Sociological Bulletin*, Vol. 67, Issue 3, pp. 334–347. Copyright 2018 © Indian Sociological Society. All rights reserved. Reproduced with the permission of the copyright holders and the publishers, SAGE Publications India Pvt. Ltd, New Delhi.

2 On Sunday, 17 January 2016, Rohith Vemula, a research scholar in Science Technology and Society Studies at the University of Hyderabad, committed suicide. Rohith was Dalit. I dedicate this chapter to Rohith, a student of science, technology and society studies of the University of Hyderabad. His suicide note articulates a love for science, for nature and for people. An- other, a di erent, an 'alternative' love, a love without hurt, a love, which according to Martha Nussbaum must matter to justice, and to nations that build themselves on hate, a love, which Hilary Rose argues, science – in its quest for power and control – forgets.

3 A necessary caveat here is the discussion on metaphor as material for sociological analysis. On the one hand, conventional history of science while accepting the use of the gendered metaphor in early conceptions of science such as in the works of Francis Bacon regards this as 'only language' and makes it irrelevant or redundant to ontological content. On the other, the mechanistic metaphor, often used in the context of Newtonian science, is seen as a heuristic device used to convey the meaning of the 'new Newtonian science'. This double speak on the analysis of metaphor is instructive to feminist historians of science.

4 Which was different from the Platonic view.

5 Sankhya, a classical system of Indian philosophy is explicitly dualistic.

6 In fact Bagchi traces this back to an earlier Bengal. I would like to state that though Bagchi does a Bengal specific analysis, we can apply this across various regions in India. Shankara's revival of Hinduism heavily relies on the Goddess/Mother.

References

Bagchi, J. (1990) Representing nationalism ideology of motherhood in colonial Bengal. *The Economic & Political Weekly, 25*(42–43), WS 65–71.

Bagchi, J. (2010) Representing Nationalism Ideology of Motherhood in Colonial Bengal, in Motherhood in India: *Glorifi cation without Empowerment* (Krishnaraj, M., ed.), New Delhi and Abingdon: Routledge.

Dasgupta, S. (1999) *Jagadis Chandra Bose and the Indian Response to Western Science*, New Delhi: Oxford University Press.

Fee, E. (1981) Is Feminism a threat to scientific objectivity?, International Journal of Women's Studies 4, 386.

Harding, S. (1996) *Science Question in Feminism*, Ithaca: Cornell University Press.

Krishnaraj, M. (2010) *Motherhood in India: Glorification without Empowerment?* New Delhi: Routledge.

Lloyd, G. (1996) Reason, Science and the Domination of Matter, in *Feminism and Science* (Keller, E.F. and Longino, H.E., eds.), Oxford and New York: Oxford University Press.

Mazumdar, M. (2017) Science and the Making of New Nationalist Masculinity in Colonial Bengal, in *Feminists & Science: Volume 2: Critiques & Changing*

Perspectives in India (Krishna, S. and Chadha, G., eds.), pp. 175–206. Kolkata: Stree.
Merchant, C. (1980) *The Death of Nature: Women, Ecology, and the Scientific Revolution*, New York: Harper Collins.
Nandy, A. (1980) *Alternative Sciences: Creativity and Authenticity in Two Indian Scientists*, New Delhi: Allied Press.
Nandy, A. (1989) *Science, Hegemony and Violence*, New Delhi: Oxford University Press.
Ramaswamy, S. (ed.) (2003) *Beyond Appearances? Visual Practices and Ideologies in Modern India*, New Delhi: Sage Publications.
Rangarajan, M. (2015) *Nature and Nation: Essays on Environmental History*, New Delhi: Permanent Black.

4

BUILDING 'INDIA'S FUTURE POWERHOUSE'

Discourses of 'development' and popular resistance in Northeast India

Chandan Kumar Sharma[1]

The region

The northeastern region of India, constituted of eight states, can be geographically divided into the plains and the hills. The most populous part of the region is the narrow strip of the Brahmaputra Valley of Assam. One of the most rain-fed areas in the world, the region is ecologically sensitive, geologically fragile and is a global biodiversity hotspot. It is seismically one of the most active regions in the world and has experienced a number of major earthquakes. The last two major earthquakes in 1897 and 1950, both measuring above 8.7 at the Richter scale, caused considerable geographical upheaval in the region.

The northeast region is endowed with considerable water resources. Most of these rivers are constituents of the Brahmaputra river system, another river system being the Barak river system. The Brahmaputra, one of the mightiest rivers in the world, is an international river which originates in Tibet and merges in the Bay of Bengal after flowing through Assam and Bangladesh. This 2,880-km-long river carries the second largest sediment yield in the world, while it ranks fourth in terms of water discharge.[2] The Brahmaputra river basin includes four countries: China, India, Bhutan and Bangladesh. Throughout its long course, the Brahmaputra receives numerous tributaries with their own idiosyncrasies. Therefore, the significance of the Brahmaputra can be understood only together with its tributaries – as a river system. The Brahmaputra river system along with its major tributaries in AP, such as the Siang, Dibang, Lohit, Subansiri, Kameng, etc., is not only critical for the ecology and livelihood of the indigenous communities of AP but is also inextricably linked with the floodplain ecology of wetlands and grasslands in the Brahmaputra Valley. These linkages, for example, are clearly visible in world-renowned ecosystems such as Kaziranga, Manas and Dibru-Saikhowa National Parks in Assam.

The economy of all the northeastern states is primarily agrarian. An overwhelming majority of the population is dependent on natural-resource-based livelihoods. Shifting cultivation (*jhum*) is the dominant traditional system of land use in the hills and plays a critical role in maintaining agricultural biodiversity and providing food security. The diversity of communities is also accompanied by their unique socio-cultural, agro-ecological and land tenure systems. The Brahmaputra river system is the lifeline for livelihoods such as fishing and agriculture of local communities in its floodplains. The unique ecology of the river determines the livelihood patterns, customs, food habits, music, religious beliefs, etc., of the indigenous communities of the region. Briefly speaking, the very socio-cultural identity of the indigenous people of the valley is intertwined with this river system. Similarly, the Barak river system is extremely critical for the life and livelihood of the people in the Imphal Valley in Manipur, the Barak Valley districts in Assam and for those living along its riparian areas of Bangladesh.

A security frontier

It is to be noted that the northeastern region, located in the northeast corner of India, is not a mere geographical entity; it is also a security frontier.[3] After the British annexed the region to the British Indian territory in the 19th century, it turned into a security buffer for the Indian 'mainland' vis-à-vis China and Burma. After India's independence, the creation of (east) Pakistan (later Bangladesh) carved out of the British India and then the Chinese invasion of 1962 further perpetuated that security paradigm. State policies with regard to the region have also been informed by it ever since. This largely seems to account for the indifference of the Government of India (GoI) in making big investments in the region, which is always under threat from external powers. Nevertheless, some observers of the region point out that since the 1980s the region has witnessed a shift in the GoI's approach towards it, i.e., from a 'security paradigm' to a 'development paradigm', leading to a considerably increased amount of government expenditure in the infrastructure sector.[4]

This policy was pushed further from the late 1990s. The GoI, in tune with its neoliberal economic policies adopted since the early 1990s, began considering heavy investments of both government as well as private capital in the region. This shift appears to be a result of three important factors: the emerging financial capability of Indian public and private sectors; the rising needs of electricity in the industries and cities of 'mainland' India; and the growing recognition of the Indian state about the need to assert its presence in the region, especially in AP, on which China lays its claim as its territory. The Indian state has already been able to make some penetration in AP in the political domain through various political institutions and agencies and in the cultural domain through expansion of education, Hinduism and Hindi. However, now it seeks to assert its physical presence in the region. This project of the Indian state in the region can be described as what political scientist Sanjib Baruah calls the push for "nationalizing space".[5] Historian Bérénice Guyot-Réchard, however, describes this process as an attempt at state-making rather than nation-building. She argues that developmentalist policies of the Indian state in the region often undermined the possibilities for nation-building.[6] The recent big push for building infrastructure such as roads, bridges and hydel projects in AP is but a manifestation of the Indian state's new developmentalism which, contrary to Ramesh (2005), seems to be deeply informed by security concerns.

Hydel power and popular imagination in the region

The perennial rivers in the region have been the subject of imagination as a potential source of development since independence, especially in the Brahmaputra Valley of Assam, the most populous and developed part in the region. This imagination is expressed not only in political rhetoric but also in the cultural domain. Several songs of Bhupen Hazarika, the legendary bard and cultural icon of Assam, are dedicated to the celebration of the greatness of these rivers, especially that of the Brahmaputra. Some of his songs also express ire at the vagaries these rivers cause and a resolve to control them for the benefit of the people by using modern technology.[7] Through these songs, Hazarika merely represented the contemporary Assamese nationalist imagination. Using these rivers for generating power through hydel projects with provisions for irrigation and flood control became part of this imagination, which found strong political expression in the 1980s. The first regional party government in Assam under Asom Gana Parishad (AGP), which came to power in the mid-1980s, submitted a memorandum to Rajiv Gandhi, the then prime minister of India, on 14 September 1988 demanding that a scheme prepared by the Brahmaputra Board[8] for several multipurpose hydel projects be implemented. It also expressed the view that of these projects the two on Dihang and Subansiri (major tributaries of the Brahmaputra in AP) would particularly reduce the intensity of flood and at the same time generate cheap hydroelectric power to feed the national grids in the entire northern belt up to Uttar Pradesh. Interestingly, this proposal of AGP was opposed by the AP government at the time, which was reiterated again in 2005.[9]

This was indeed in stark contrast to what was to happen after a decade and a half when the Assamese public opinion turned against dam building in the region. Before this turned into state-wide protest, apprehensions and protests were expressed at the grassroots, in the areas like Bihpuria in the Lakhimpur district close to the Assam-AP border, against the proposed 405 Megawatt (MW) Hydel Project on Ranganadi river. This project situated in the upstream of the river in AP was commissioned in 2002. While it was asserted by the government that the Ranganadi Hydel Project will mitigate flood, the apprehension of the downstream local communities came true when the released excess water of the Ranganadi dam during the peak of summer led to flash floods in North Lakhimpur town, the headquarters of the Lakhimpur district, and large areas around it on 14 June 2008. This was a shocking but eye-opening incidence to many, who were either indifferent to the issue or were advocates of the flood control capacity of dams. Since then flash floods from the Ranganadi project have become a recurrent phenomenon in every monsoon.[10]

India's 'future powerhouse' and the Lower Subansiri Hydel Project (LSHEP)

In August 1998, the GoI announced its Policy on Hydro Power Development, followed by 50,000 MW hydroelectric initiatives in May 2003. Both of them had their major focus on the hydroelectric potential of the Brahmaputra basin. The northeast (especially AP) was identified as India's future powerhouse with the GoI aiming to generate 63,000 MW of hydel power by building as many as 168 dams on almost all its perennial rivers. The news was received with mixed reaction.

Early 2001, the 2,000 MW LSHEP under the National Hydroelectric Power Corporation (NHPC) was in the news. The project was to come up at a site near Gerukamukh at the Assam-AP foothills in the lower part of the river Subansiri. The project made the downstream communities of Assam apprehensive for several reasons. Firstly, there was an effort to begin construction of the project without any public hearing for environmental clearance which is legally mandatory. When some civil society organisations raised an outcry about this, a public hearing for the project took place on 4 September 2001. The Assam State Pollution Control Board got the local Assamese and Mising communities, mostly semi-literate, to sign a resolution endorsing the project after a presentation was made in English and Hindi.[11] This was opposed by some present at the hearing and downstream concerns were raised. This, however, was ignored. The clearance for use of forest land for the project in a biodiversity-rich area (covering the adjacent Subansiri and Dulung Reserve Forests) was already submitted by the Government of Assam (GoA) to the GoI without any proper study.

Besides, several violations of environment and forest laws as well as of the Supreme Court of India (SC) orders have been reported in the subsequent years pertaining to the construction of the LSHEP. For example, in April 2004 the SC imposed conditions that

> the NHPC would ensure that there is no siltation down the Subansiri during the construction phase; b) under no circumstances,

> the excavated material would be dumped either in the river or any other part of the National Park/Sanctuary or the surrounding forests. Despite this, there has been indiscriminate dumping of muck and debris in the river, repeatedly recorded and reported by the local communities since 2004.[12]

Even the meaning of 'downstream' has been arbitrarily defined. The Union Ministry of Environment and Forests (MoEF) fixed a 10 km distance from the powerhouse for evaluating the downstream impacts of large hydel projects,[13] discounting not only the concerns of the communities in the Brahmaputra Valley but even their *locus standi* on the issue. That these non-transparent procedures were often in violation of the established legal provisions only contributed to the already unfolding public speculations in the Brahmaputra Valley that LSHEP was not to cater to the interests of the region. Civil society organisations in Assam such as All Assam Students Union (AASU) and Asom Jatiyatabadi Yuba Chatra Parishad (AJYCP) along with several environmental NGOs became vociferous against the NHPC.

What is interesting, as indicated earlier, is that the AP government had warned the NHPC against undertaking the construction of LSHEP on at least two occasions in 2005. In two letters in January and March of that year addressed to the Chairman and Managing Director of the NHPC, the Power Secretary of AP had pointed out 'serious procedural lapses' stating that its approval had not been acquired for the project. The letters held that dam construction activities 'may aggravate the prevailing boundary dispute between the people of AP and Assam'.[14] The letters had further noted that displacement of people and their relief and rehabilitation are 'very difficult issues'.[15] They had alluded to the previously mentioned SC judgment of April 2004 and held that the implementation of the project was in violation of the SC order and 'therefore tantamount to contempt of the SC'.[16] However, the AP government soon did a *volte face* and became a champion of big hydel projects in the state.

Influence of external factors

The emerging public apprehensions about dams in Assam were informed by the developments around the Narmada Bachao Andolan (NBA) throughout the 1990s in resistance to the Sardar Sarovar Dam on the river Narmada.[17] The NBA generated new debates and consciousness in India on the criticality of environmental issues and created a new narrative against the conventional state-driven (or in conjunction with big capital) development discourse where the voice of the common people (especially, the tribal and marginal communities) gets submerged under the burden of 'national' or 'public' interest.

Meanwhile, the report by the World Commission on Dams (WCD) which had come out in 2000 received public attention in Assam. The WCD was constituted by the World Bank and the International Union for Conservation of Nature (IUCN) in 1998. Although there have been studies and opinions on the adverse impact of dams, especially big dams, even prior to the WCD report, the latter was the first worldwide independent study conducted by international experts. It said,

> [W]hile dams have made an important contribution to human development and benefits derived from them have been considerable, in too many cases an unacceptable and often unnecessary price has been paid to secure these benefits, especially in social and environmental terms, by people displaced, by communities downstream, by taxpayers and the natural environment.[18]

It further said that since the 1970s there has been a significant decline over construction of dams in Europe and North America and that a process of river restoration by decommissioning the dams has found a new priority.[19] The report brought about a significant change in the worldwide public outlook toward big dams and even many of its erstwhile votaries turned sceptics if not outright against it. Although the NBA-led anti-dam movement in the Narmada Valley already heralded the anti-dam movement in India, the WCD report strengthened the anti-dam voices in the country considerably.

Further, the major earthquake of October 1991 that devastated the Garhwal region of the state of Uttarakhand and the apprehension about the threat to the Tehri dam in the region were also a talking point in Assam. Subsequently, the seismic threat to the dams proposed in AP, which shares geo-morphological characteristics with Uttarakhand, became a key issue for the anti-dam protesters. It is to be noted that the 1950 earthquake and the devastation it caused are still fresh in the minds of many both in Assam and AP. Further, the fact that the region is ecologically most sensitive and recognised as a global biodiversity hotspot also came to inform the apprehensions about dam building in the region. The increasing opposition of many environmentalists, geologists, engineers, social scientists and activists in Assam in subsequent years against dam building in AP has to be understood in this context. They express serious apprehension about the impact of the dams not only in the downstream areas of Assam but also in the ecologically sensitive upstream areas of AP.

However, people in the northeastern region are not new to the pitfalls of dams and to anti-dam protests. The construction of the Kaptai dam in the Chittagong Hill Tracts of Bangladesh in the 1960s submerged the traditional homelands of the Hajong and Chakma communities and forced thousands of them to migrate into various parts of northeastern India, especially AP. In the 1970s, the Gumti dam in Tripura led to the submergence of a huge swathe of tribal land. The Loktak hydel project commissioned in the 1980s seriously impacted the wetland ecology of the Loktak Lake in Manipur. Similarly, the impending loss of home, land and livelihood has led to many years of opposition to the Pagladiya project in Lower Assam and the Tipaimukh project on the river Barak in Manipur along its border with Mizoram. In fact, the movement against the 1,500 MW Tipaimukh hydel project located within a distance of 100 km from the Bangladesh border attracted much wider attention not only because it poses a serious threat to the livelihood of thousands of people in Manipur but also because the project has been opposed by Bangladesh for its possible adverse impact on the flow of the river Barak into that country. The Bangladeshi experts complain that the dam would jeopardise the livelihood of the communities including peasants and fisherfolk living in the lower riparian areas of the Barak in Bangladesh.[20]

Again within Assam, during the rainy season, the districts of Marigaon and parts of Nagaon in central Assam are regularly flooded by the excess water released by dams such as the Karbi-Langpi, Kopili and Umtru with capacity from 50 MW to 100 MW. Similarly, the Kurichu dam in Bhutan with a capacity of only 60 MW has been causing considerable periodic destruction of crops and other resources in the lower Assam districts of Barpeta, Baska, Nalbari and Kamrup. In upper Assam too, the excess water released from the 75 MW Doyang hydel project in Nagaland has caused significant damage in the downstream areas of Golaghat district of upper Assam in recent times. The case of Ranganadi project in Lakhimpur district upper Assam is already mentioned. The intensity of this destruction appears to be only increasing with every passing year. In the absence any alternative livelihood opportunities other than agriculture, a large section of youth from these districts has migrated to other states of India as daily wage earners in recent years.[21]

In the background of all these, the policy of the GoI to turn the northeast into 'India's future powerhouse' led to widespread apprehension and disquiet in the region, especially Assam. Assam became the epicentre of anti-dam protests because of the simple fact that most of these proposed dams are to come up in the rivers in AP which is located in Assam's upstream, and therefore Assam is destined to suffer from a range of detrimental downstream impacts of these hydro projects. These protests are based on the apprehension that this 'development' initiative of the government would spell disaster to the river ecosystem and the livelihood and cultural heritage of millions of people of the region. The protests are also informed by the fear that the region is highly seismic and geologically fragile and there is always a possibility of a dam break leading to a major catastrophe in the downstream.

Thus the extant public imagination about the great development potential of the Brahmaputra and its tributaries took a turnaround and anti-dam protests assumed a bigger dimension. The advocates of big dams tried to convince the people by arguing that the hydel projects in AP would be 'run of the river' and therefore would be environmentally benign. But environmentalists have pointed out that this is not true. For example, the reservoir of the LSHEP will submerge a 47 km length of the Subansiri river. This will also cause drastic daily fluctuation in river flows downstream due to power generation patterns, particularly in winter. For example, the average winter (lean season) flow in the Subansiri in its natural state is approximately 400 cumecs. Both the ecology of the downstream areas and people's use of the riverine tracts in winter is adapted to this 'lean' but relatively uniform flow of water on any particular day. *Chapories* (riverine islands and tracts), for example, which are exposed and drier in winter, are used for both agriculture and cattle grazing purposes, by local communities and simultaneously by wildlife. Similarly, in summer, local communities are engaged in fishing, collection of driftwood, ferrying of passengers by boat, etc. which are important sources of livelihood for them. After the commissioning of the LSHEP, flows in the Subansiri in winter will fluctuate drastically on a daily basis from 6 cumecs for around 20 hours (when water is being stored behind the dam) to 2,560 cumecs for around four hours when the water is released for power generation at the time of peak power demand in the evening hours. Thus, the river will be starved for 20 hours and then flooded for four hours with flows fluctuating between 2 per cent and 600 per cent of normal flows on a daily basis![22]

Thus, various downstream impact concerns have been raised by the people of Assam and other parts of the region. These include loss of fisheries; changes in *beel* (wetland) ecology in the flood plains; impacts on agriculture on the *chapori*es; impacts on various other livelihoods due to blockage of rivers by dams; increased flood vulnerability due to massive boulder extraction from river beds for dam construction and sudden water releases from reservoirs in the monsoons; and dam safety and associated risks in this geologically fragile and seismically active region. It was around this time that the previously mentioned Ranganadi flash flood of June 2008 occurred and gave an immediate fillip to the anti-dam movement.

At this juncture, in another turnaround, the Congress party-led government of AP began showing much enthusiasm to the initiative of the GoI toward building dams in the rivers in the state. It began signing dozens of memorandums of understanding (MoU) with various dam developers at a frenetic pace while the people of Assam kept crying foul and demanding the GoA to take up the matter urgently with the governments of India and AP. However, the GoA remained silent on the matter.

Significantly, each of these MoUs had been accompanied by huge monetary advances taken from project developers at the time of inking the deal, before any public consultations were done and clearances obtained. "This kind of process of signing MoUs, where monetary advances are paid up front, greatly compromises the manner in which subsequent clearances take place, as projects are considered as a *fait accompli* by both the developer and the state government."[23]

The vacillation of the political parties to the evolving scenario was striking. While the Congress-led state government in Assam was in all agreement with the Congress-led GoI on the issue of dam building, the main opposition party AGP, which once demanded a multi-purpose project in Subansiri, remained indecisive until 2010 and came out with vague statements regarding the construction of the Subansiri dam. The other political parties in Assam, such as the Bharatiya Janata Party (BJP) and All India United Democratic Front (AIUDF), also took an ambiguous position until mid-2010.[24]

Meanwhile, a perception that the fate of LSHEP would decide the course of anti-dam movement in the region animated the public protests against the dam. It is to be noted that the LSHEP is the first dam proposed to be built in AP under the new hydro power policy of the GoI and it thus became the focal point of the anti-dam protests. While AASU and AJYCP were giving leadership to the popular anti-dam voices in Assam, the entry of Krishak Mukti Sangram Samiti (KMSS) into the scene in the year 2006–2007 radicalised the anti-dam movement in the state. A farmers' body, KMSS emerged as a powerful social organisation in the middle of the last decade, leading a series of mobilisations articulating the interests of the marginal groups.[25]

Public uproar in the upstream

This had its impact, though relatively small, in Arunachal too. While some civil society groups and opposition political leaders in the state had been already voicing anti-dam sentiments, they were rather isolated and not given much importance. However, KMSS-led mass mobilisations had its own stimulating effect on the anti-dam mobilisations in Arunachal.

One of the major arguments to build large hydel projects in the northeast is that there is relatively 'small displacement' of people on account of its low density of population as compared to other parts of the country. But considering other parameters of this sparsely populated landscape, this logic of 'small displacement' appears rather problematic. For example, *jhum* cultivation covers a large tract of land and the taking over of the land for hydel projects and other related activities will create increasing pressure on land. It would shorten the *jhum* cycle and enhance the pressure on the surrounding areas, adversely affecting not only the livelihoods of *jhum*-dependent communities but also the environment.[26] Further, Mite Lingi, the Chairman of the Idu Indigenous People's Forum contends that the 'small displacement' argument to sell these so-called benign projects needs to be confronted. The entire population of the Idu Mishmi tribe is around 9,500 and by this faulty logic the series of large hydel projects planned in their habitat, the Dibang Valley, will have little social impact even if their entire population were displaced.[27]

The concerns being expressed are not restricted to the issue of displacement alone. The 1,750 MW Demwe Lower Project in the Lohit district on the river Lohit poses to jeopardise the very existence of the Parasuram Kunda, a much revered pilgrimage site for millions of Hindus. The site is also deeply associated with the myths and legends of the local communities such as the Mishmis, Deoris, Chutias, etc. Besides its significance as a religious and tourist spot, the site is also ecologically sensitive.

Another major concern for the people in AP has been the influx of thousands of immigrant labourers for construction of the dams. These are long-term projects, and the immigrant labourers are poised to settle in the dam sites for long periods, if not permanently, posing a serious threat to the existing demographic structure in the hills, where most of the communities are numerically very small. Mite Lingi further adds,

> We have been given constitutional and legal protection, particularly with respect to our land rights and restricted entry of outsiders.[28] 13 large projects in the Dibang Valley will bring in outside labour upwards of 100,000 people for long periods as these are long gestation projects. We are concerned about the demographic changes and other socio-cultural impacts associated with this as the Idu Mishmis are . . . (very small) in number. The development policies are a glaring contradiction to the constitutional and legal protection we have been given.[29]

It may be mentioned that the 3,000 MW Dibang dam will alone result in submergence of an area of 3,564 hectares. The land acquisition for the project is going to displace 115 families in five villages and affect 744 families in another 39 villages.[30]

Engagement with the government

Meanwhile, amidst public protests against the LSHEP, the Power Minister of Assam after a consultation with the dam builder NHPC and AASU, which was opposing the dam, appointed an expert committee for the downstream impact assessment of the project in 2008. The Assam government maintained that it would stick to the expert committee report on the dam. Around the same time, on demand from some opposition political parties of Assam, a committee with members of the Assam Legislative Assembly from the ruling as well as the opposition parties was also constituted to look into the issue. Both the committees in their interim reports asked the NHPC to stop the dam until the committees submitted their final reports. But neither the NHPC nor the GoA paid any importance to the interim reports. After the final report of the expert committee was submitted in August 2010, which raised questions about the selection of the dam site and its downstream impact, the GoA changed its earlier stand and pressed for an 'international expert' panel to study the downstream impact of the dam.

This set off a debate on the eligibility of experts to the study of dams which Sanjib Baruah termed as 'the politics of expertise'.[31] While the report vindicated the stand of the civil society groups, the attitude of the government and the dam builders created further resentment among the people.

Interestingly, during the same time, the MoEF stalled the construction work of a number of dams and other projects on ecological and cultural grounds in different parts of the country. For example, the case of the three dams on River Bhagirathi may be cited where GoI scrapped the construction of Lohari Nagpala, Pala Maneri and Bhairon Ghati dams in August 2010. GoI noted that Bhagirathi being a tributary of the Ganga, construction of dams on the river would restrict the flow of water to the latter, endangering the river considered holy in Hinduism. GoI also declared a 135 km Ganga river stretch from Gaumukh to Uttarkashi in Uttarakhand an eco-sensitive zone.[32]

However, the anti-dam protesters in the northeast became indignant that no such sensitivity was shown in case of the dams in the region. In the wake of the rising anti-dam protests, the then Union Minister of Environment and Forests Jairam Ramesh came to Assam on 10 September 2010 and heard the opinion of various civil society organisations from various states of the northeastern region in a public meeting in Guwahati with respect to the proposed and under-construction hydel dams in the region. The civil society organisations were unequivocal in their opposition to the dams. The minister prepared a report reflecting the prevailing public opinion in the region and submitted the same to the PM, which ended with the warning that the "feeling in vocal sections of Assam's society particularly appears to be that 'mainland India' is exploiting the North-East hydel resources for its benefits".[33]

In the meantime, in Assam, as the elections to the State Assembly neared, the opposition political parties discarded their earlier ambiguity and became somewhat more vocal against the LSHEP. AGP, the main opposition party, came out openly against it, citing recent experiences and knowledge on the adverse impact of dams. On this question, a whole day was devoted to a debate in the Assam State Assembly. However, the defeat of the opposition in the assembly elections held in April 2011 was hailed by the ruling Congress party as their moral victory on the issue.

That could not dampen the public protest. The protesters demanded that the concerns raised in the expert committee report on the LSHEP must be addressed and that a basin-wide cumulative impact assessment of various dams proposed to be built in AP be undertaken before going ahead with their construction. However, the government's dilly-dallying on this only aggravated popular resentment. The protesters, led by organisations such as KMSS, Takam Mising Poring Kebang (All Mising Students' Union) and AJYCP blocked National Highway 52 for many days to stop movements of all materials to the dam site at Gerukamukh. This caused all activities to stop in the construction site since December 2011. Although NHPC maintained that they took all necessary measures to ensure the safety of the dam and that the apprehensions raised by the expert committee were unfounded, the members of the expert committee rejected NHPC claims.[34]

Nevertheless, following Ramesh's public consultation in Guwahati, a two-member technical committee under C. D. Thatte was constituted by the Planning Commission in 2011 under the instructions of the prime minister's office, to recommend how NHPC could move forward with the construction of the LSHEP. The committee submitted its report in July 2012 which included a number of critical comments and questions about various aspects of the project, which have been already articulated by the anti-dam protesters.[35] The report of the committee thus further vindicated the apprehensions of the protesters.

Disquiet over other dams

However, such developments could not deter the government from going ahead with more other dams in Arunachal. Among them, the previously mentioned 1,750 MW Demwe Lower Project coming up on the river Lohit and the 2,700 MW Lower Siang dam on the river Siang have become sources of much controversy. The former, which could jeopardise the sacred site of Parasuram Kund, has faced resistance from the local people and environmental activists. It may be noted that India is a signatory to the Akwé: Kon voluntary guidelines of 2004 for the conduct of cultural, environmental and social impact assessment regarding developments proposed to take place on, or which are likely to impact on, sacred sites and on lands and waters traditionally occupied or used by indigenous and local communities.[36] But its stand on the Demwe Lower Project clearly contradicts the spirit of the Akwé: Kon guidelines. Similarly, the local Adi community has been up in arms against various hydel projects on the river Siyang in the East Siyang district of AP since 2010, besides a host of other hydel projects on the rivers in the Siyang Valley. The protest was especially targeted against the 2,700 MW Lower Siyang Hydropower Project.[37] The protesters, led by organisations like Forum for Siyang Dialogue, argue that these projects would destroy the livelihood and culture of the Adi community living in the valley.[38] In April 2012, the government had to cancel a number of public hearings in the East Siang and West Siang districts of Arunachal after anti-dam protesters resorted to violence. The protesters not only destroyed public properties, they also attacked the houses and properties of local supporters of dam.[39] The leaders of the Siang Bachao Federation issued an ultimatum to all dam builders to immediately leave the Siang Valley or face serious consequences.[40]

In case of the 3,000 MW Dibang dam, even before the foundation stone of the India's largest project was laid on 31 January 2008 by the prime minister of India, the project faced resistance from the local Idu Mishmi tribe for alleged violation of India's environment and forest clearance processes. From 2007 to 2011, the Lower Dibang Valley district, especially its headquarters Roing town, was the hotbed of public protests. There were police excesses and in a police firing in October 2011 eight school children were wounded. Although the state administration tried to downplay the reasons behind the protests by portraying them as a result of Maoist instigation,[41] the public concerns had been validated by the Forest Advisory Committee of the MoEF, which had turned down this NHPC project twice in 2013 and 2014, respectively.[42]

Further, the strategically sensitive Tawang, located close to the China border and an Eastern Himalayan biodiversity hotspot, witnessed protest against 13 proposed dams on the rivers in the Tawang Valley to develop 3,500 MW of hydropower. These dams are on rivers that are tributaries of the river Manas, which is a major tributary of the Brahmaputra. Since early 2012, hundreds of the members of the local Monpa tribe and Buddhist monks of this renowned pilgrimage town rose in protest against the proposed construction of 13 dams in the Tawang Valley. They staged a mass rally against dams in the district in April 2012. In September 2012, there was another anti-dam protest rally. In December 2012, protests turned violent when anti- and pro-dam activists clashed with each other. Several men and women activists of the Save Mon Region Federation (SMRF), one of the outfits spearheading the protests, were arrested on 23 December 2012 for allegedly ransacking the NHPC office at Tawang. The protesters expressed apprehension that these dams would destroy not only the local culture but also the local ecology, which houses many locally threatened and endangered plants and animals including the clouded leopard, snow leopard, red panda and Macau mountain goat. Lobsang Gyatso, a Lama at the Tawang monastery and general secretary of SMRF, says that most of these projects are unnecessary. He says, "the population of Tawang is about 49,000. We don't need so many hydel projects to meet the electricity demand of our people. Small hydro-projects would suffice. All these large dams are meant to generate electricity to be sold outside, at the cost of our livelihoods and ecology."[43] In 2012, Lamas under the SMRF's banner met the then prime minister Manmohan Singh and power minister Sushil Kumar Shinde to scrap the hydel projects in the area. But nothing came out of the meetings.[44]

Ignoring all protests and apprehensions, the 800 MW project on Tawang Chu river was cleared at a meeting in January 2013. This happened within days after China announced three new hydel projects on the Brahmaputra in Tibet with the MoEF waiving the cumulative impact assessment for stage one clearance which it earlier insisted on. The ministry decided to delink the cumulative study on the Tawang basin from stage one forest clearance, a proposal which was also backed by the state government.[45]

After the new BJP-led Union government assumed power in Delhi in May 2014, it adopted a policy to fast track the necessary forest and environmental clearance process of many development projects throughout the country. Under this policy, the Dibang hydel project was cleared in September 2014[46] which led local activists to accuse the GoI of bulldozing local concerns. They accused the dam builders of corrupting the small local communities with promise of financial compensation. That is why, they alleged, there was no visible public protest after the clearance of Dibang hydel project.[47] It has been pointed out that the dam will submerge a large biodiversity-rich forest area with several endemic species in AP,[48] and will entail severe impact in downstream areas, primarily in Dibru Saikhowa National Park, a rich biodiversity hotspot in Assam.[49] However, six years after its foundation stone was laid and twice denied environmental clearance, the 3,000 MW dam was cleared, subject to a reduction in the dam height by 20 metres from the originally envisaged 288 metres. This clearance for India's largest hydro project and the world's tallest concrete gravity dam came allegedly after a letter on 3 September 2014, from the prime minister's office to the secretary of MoEF to 'clear the project expeditiously'. Incidentally, of the six Forest Advisory Committee members who cleared the project, four were also part of the panel that had unanimously rejected the project in April 2014.[50]

In a report submitted by Asad Rahmani, Director of the Bombay Natural History Society, who was a part of a two-member team constituted by the standing committee of the National Board of Wild Life (NBWL) in 2014 to assess the possible impacts of the 1,750 MW Demwe Lower Project on wildlife, highlighted that the project will have serious ecological implications on the Kamlang Wildlife Sanctuary and the Dibru Saikhowa National Park located in its vicinity. It would submerge areas significant for medicinal plants besides adversely affecting the migration of fish such as the golden mahseer and the habitat of the Ganges river dolphin. The report concluded that the project raises "serious concerns about downstream impacts on wildlife habitat and species. In this scenario, the precautionary principle should be applied and under no circumstances should wildlife clearance be given based on current impact assessment reports."[51] However, the second member of the committee, the Chief Conservator of Forest (wildlife) of AP, expectedly opined that more comprehensive studies were required to draw any conclusion on the downstream impacts of the project and that these studies can be done simultaneously along with the progress of the project. The state wildlife officials denied that the project will cause any significant impact on the wildlife. Arguing that the project is important for the country's strategic and development interests, they emphasised that the project be cleared soon.[52] However, no comprehensive studies mentioned previously have been witnessed.

This was in stark contrast to what was proclaimed during campaigns in Assam for the May 2014 Indian Parliamentary elections, in which barring the ruling Congress party, all other political parties adopted an anti-big-dam stand. Top leaders of BJP, the main opposition party at the time, including the present prime minister of India, Narendra Modi, during their campaigns unequivocally pitched for small hydel projects without causing ecological damage.[53] Rajnath Singh, the then president of the party, even went to a location near the Lower Subansiri dam site and addressed a huge public meeting in which he spewed fire against big dams in the region.[54]

In an interesting development, the Expert Appraisal Committee (EAC) for River Valley and Hydroelectric Projects of the Ministry of Environment, Forest and Climate Change (MoEFCC, the new nomenclature of the MoEF) met on 3 June 2015 and rejected three hydel projects of 96 MW each on the Tawang river basin: the Mago Chu, Nyukcharong Chu and New Melling hydel projects. Citing its reasons for not giving environmental clearance for the projects, the appraisal committee noted that the "Tawang sub-basin study is yet to be examined and accepted by the ministry".[55] The committee also noted that the proposal for a grant of environmental clearance needed to be looked into by the ministry. The SMRF leader Lobsang Gyatso welcomed the committee decision and said that they were not opposing all projects but emphasised that the local people should be consulted while implementing these projects. He also claimed that politicians from the area were putting pressure on the people in attempts to forcefully take land away from them.[56] However, what prompted the MoEFCC to refuse clearance to the dam remained unclear, especially when similar considerations are not observed in the case of other projects.

On the eve of a crucial meeting of the EAC to be held on 24–25 August 2015, the SMRF made an appeal to it to reject the 780 MW Nyamjang Chu project in its current form and also to hold public consultations with the locals prior to accepting the recommendations of the Tawang River Basin Study. The organisation said in its letter to the EAC members that the barrage of the Nyamjang Chu project was proposed to be located right in the middle of a wintering site of the black-necked crane, which is both ecologically and culturally important. However, *The Arunachal Times*, the popular daily from AP, reported that the EAC recommended clearance of an alarming number of hydro power projects in AP. The newspaper reported that the EAC accepted recommendation to build 26 out of 28 hydropower projects on the Subansiri river basin, although its study was not even listed among the 14 subjects that were placed for discussion. This was also done without any public consultations in AP or Assam. The committee further recommended clearances to a number of hydel projects in the Tawang Valley including Nyamjang Chu and three projects namely, Nykcharong Chu, Mago Chu and New Melling (90 MW) subject to the statutory clearance of the individual projects.[57] As mentioned above, the last three projects were rejected barely three months ago by the EAC.[58]

Meanwhile, on 30 July 2015 the National Green Tribunal (NGT) admitted an appeal challenging the environmental clearance given to the Dibang project. The MoEFCC in its environment clearance letter issued on 19 May 2015 had stated that a study shall be undertaken regarding the impact of the project on environment and downstream ecology after five years of commissioning the project.[59]

In a decision hailed as 'rare' and 'exceptional', the southern bench of the NGT suspended the forest clearance of the controversial Demwe Lower Hydel Project on 24 October 2017 hearing an appeal against the stage one forest clearance of the project on the grounds that the clearance was not given after due procedure. According to an SC order, all projects within 10 km of national parks and sanctuaries need clearance from the standing committee of the National Board for Wildlife (NBWL). In case of the Demwe Lower project, eight non-official members of the 12-member NBWL standing committee rejected the proposal. But finally, the clearance was given by the minister which was declared as illegal.[60] However, this sense of triumph for the environmentalists and activists was short-lived as NBWL cleared the project in September 2018, based on a report described as seriously flawed by the environmentalists and prepared by Wildlife Institute of India (WII).[61]

In another development, hearing a case filed by an NGO, Assam Public Works (APW), the eastern zone bench of the NGT on 16 October 2017 asked the MoEFCC to constitute a three-member committee to study the LSHEP with at least one member from the northeast. It said the committee should study the proposed alternative dam structure and the reports of the earlier committees and submit its report to the tribunal in three months.[62] The bench gave a month to the ministry to form the committee. However, while the MoEFCC took more than a month to constitute the committee, it was constituted only with former government employees despite NGT's order to form a committee with accomplished experts on seismology, hydrology of rivers and river ecosystems of the Himalayas and the northeast. In April 2018, NGT accepted a petition submitted by the APW challenging this subversion of the NGT directive.[63] However, after this petition was dismissed by NGT in July 2019, APW filed a petition in the SC in November 2019 seeking a stay on NGT order.[64]

The China angle

The Chinese project of building a 510 MW Zangmu dam in Yarlong-Tsangpo (as the Brahmaputra is known in Tibet) and its alleged plans to divert its water toward the north has been a cause of major concern in the downstream Indian states of Assam and AP for several years. Naturally, communities in downstream areas of northeast India are alarmed as the damming of the Brahmaputra in Tibet may seriously affect their life and livelihood by depriving the river of its water flow in a major way, leading to devastating consequences on the ecology of its downstream areas.[65]

The pro-dam voices interestingly often use the alibi of the provision known as the 'first users' right' vis-à-vis China for constructing dams in Arunachal. They argue that by building dams on transnational rivers in AP before China does it, India can deter China from building any major projects on these rivers which might affect India's use of its water in the future. Indeed, the Helsinki Rules on the Uses of the Waters of International Rivers stated in 1966 that all bordering nations have a right to equitable shares of the water resources that cross national boundaries, except where other agreements between bordering nations exist. But the Helsinki Rules never became universal law. Repeated attempts by the United Nations failed to bring about a legally binding international treaty on the first users' right. Only on 17 August 2014, such a treaty could muster the support of the minimum 36 member countries of the United Nations. Incidentally, major countries including India and China are not yet signatories to this treaty. It clearly shows the campaign of falsehood that the advocates of big dams in AP have been engaged in.[66]

Further, what is important for the people in the northeastern region is that the proposed dams in AP will not have any less adverse impact on the region than what the Chinese dams might cause. While a section of advocates of the hydel dams in Arunachal are vociferous against the Chinese dams, they stop short of stating that the dams in the state will also have similar affect in Assam and Bangladesh.

Conclusion

It is clear that the public opinion in the northeast is quite informed with regard to the social and ecological implications of big dams. This awareness is a result of not only their exposure to anti-dam movements elsewhere and scientific studies about the adverse impacts of dams, but also from their own lived experiences. In the upstream AP, initially an impression emerged, due to government propaganda, as if the communities there extended unambiguous support to the proposed dams in the hills. But soon anti-dam protests engulfed various parts of the state once the communities realised the fallout of the scores of hydel projects proposed in their habitats. Although under persistent pressure from the government and manipulative tactics by the dam builders, the popular resistance has declined in certain areas, the protests continue.

Highlighting the impact of taking over land for various development projects on the traditional land ownership system among the tribes of AP, Tapir Gao, now the state BJP president and a Member of Parliament, argues, "not a single strip of land or water in AP is free. In accordance with the traditional laws, the local tribes own everything. Now the government is handing over land to private parties without holding any consultation with the local people."[67] Gao has in fact revealed a very sad aspect of changing land and class relations in the tribal societies of northeast India.[68] Highlighting the threat to the fragile demographic landscape of the state from migration of outsiders if the proposed hydel projects are implemented, Ojing Tasing, a leader of the All Arunachal Pradesh Students Union, states, "no one is against development. But we do not want this kind of development. It will destroy the heart and soul of Arunachal."[69] The mood in the hills is succinctly captured by Mamang Dai, an eminent writer from AP, when she says: "This development is like an invasion!"[70]

Nevertheless, it is worth noting that many issues and concerns raised by the public with regard to dam building in the northeast have been subsequently validated even by various authorities, bodies and agencies constituted by the GoI. This has undoubtedly put some restrictions on the arbitrary and non-transparent manner in which the government and the dam builders used to operate previously. Under such pressure, the NHPC has already agreed to reduce the original height of Lower Subansiri dam from 157 m to 115 m. The critical observations made by the Thatte Committee on LSHEP may also be noted in this context. Besides, the new public consciousness that the protests have generated about the social and ecological consequences of dams, and for that matter, any other development project in the region, in general cannot be undermined. It is a different matter that the achievements of the anti-dam protesters have remained much short of their goals. There is not even a semblance of recognition to their crucial demands such as a comprehensive study of the geological and hydrological features of the Brahmaputra basin and a cumulative impact assessment of all upstream dams in AP before clearance is given to their construction.

The discussion above clearly shows the unilateral and arbitrary way in which the GoI is going about the hydel projects in the northeast. There has been wanton violation of not only the popular democratic opinion but also the established procedures, norms and laws of the land. In this context, it is worthwhile to note that in a landmark case between *MC Mehta vs Union of India* (AIR 2004 SC 4033), the SC held that

> in such matters (involving actions that pose a threat of serious or irreversible damage), many a time the preferable option is not clear. If an activity is allowed to go ahead, there could be irreversible damage to the environment; if it is stopped there could be irreparable damage to the economic interests. In case of doubt, however, protection of environment should take precedence over economic interest.

However, despite the serious threats that the GoI's agenda poses to the fragile environment and the tribal communities in the region, the government seems determined to go ahead with its big dam policy in the region. Its developmentalist logic that these dams will bring prosperity has barely convinced the people of the region. It is no secret that the power generated by these dams will be used to cater to the growing needs of the power-hungry big industries and cities outside the region. What the region would get is negligible although the price it is asked to pay for this is disproportionately huge. There is a strong perception in the region that the Indian state is pushing these projects at the behest of the big capital under its neoliberal economic paradigm, that it only betrays the approach of the Indian state towards the region as a distant periphery, a security frontier. There is no gainsaying that such a perception is not warranted in a sensitive region like the northeast which has already witnessed series of protest movements, insurgencies and violence on the question of the state's exploitation of its resources. While one can hardly overemphasise the importance of development in the region and need for state intervention in this, it is also required that such a vision of development is in tune with the social, cultural, economic and physical specificities of the region. The Indian state can indeed actualise such a vision of development by initiating a creative democratic dialogue with the civil society in the region for redefining its conventional relationship with the latter.

Notes

1 This chapter is an updated version of an essay that was originally published as Chandan Kumar Sharma, 'Dam, "Development" and Popular Resistance in Northeast India', *Sociological Bulletin*, Vol. 67, Issue 3, pp. 317–333. Copyright 2018 © Indian Sociological Society. All rights reserved. Reproduced with the permission of the copyright holders and the publishers, SAGE Publications India Pvt. Ltd, New Delhi.
2 Goswami and Das (2003: 15–18).
3 Baruah (2005: 33–58).
4 Ramesh (2005: 17–71).
5 Baruah (2005: 53).
6 Guyot-Réchard (2013: 22–37).
7 In one of his songs of the 1970s on the river Kopili, a tributary of the Brahmaputra, Hazarika pays tributes to the engineering enterprises to tame the fury of the river (a eulogy to the upcoming Kopili Hydel Project), which ravages crops every monsoon, in order to produce the 'brightness of the thousand Sun'. Another song of the 1970s by his younger sibling Jayanta Hazarika recounts the tale of the changing topography of the Luit (another name of the Brahmaputra) resulting in increasing ferocity of its flood. But in the same breath, he warns the mighty river of enslavement ('*bhritya*') with modern technology.
8 The Brahmaputra Board was set up in accordance with the provisions of the Brahmaputra Board Act of 1980 to take up integrated water resources planning and management in the Brahmaputra basin in a comprehensive manner. The Board comes under the Ministry of Water Resources, government of India.
9 Baruah, 'Arunachal had warned NHPC on dam project' (www.hindustantimes.com/newdelhi/arunachal-had-warned-nhpc-on-dam-project/article1-824975.aspx).

10 The Assam Tribune, 'Dam-induced flood & nature's fury devastate North Lakhimpur' (www.assamtribune.com/scripts/detailsnew.asp?id=jul1117/state050).
11 Ahmed (2003: 30).
12 Vagholikar and Das (2010: 16).
13 Ibid.: 9.
14 Baruah, 'Arunachal had warned NHPC on dam project' (www.hindustantimes.com/newdelhi/arunachal-had-warned-nhpc-on-dam-project/article1-824975.aspx); for an understanding of the boundary dispute see Sharma (2017).
15 Ibid.
16 Ibid.
17 Baviskar (2004).
18 Report of the World Commission on Dams (2000: xxviii).
19 Ibid.: 10.
20 'Tipaimukh high dam' (www.internationalrivers.org/resources/tipaimukh-high-dam-3499); 'Tipaimukh dam to be disastrous for Bangladesh' (www.thedailystar.net/tipaimukh-dam-to-be-disastrous-for-bangladesh-23391). As of now, there is silence on the Tipaimukh dam, apparently on account of the opposition from the India-friendly Awami League government in Bangladesh. Instead, there is talk of two dams to be built by NHPC and NEEPCO in the Barak tributaries.
21 Author's presentation titled 'Socio-economic impacts of big dams in downstream areas of Assam' at the public consultation between Jairam Ramesh, the then–Minister of Environment and Forest, and civil society organisations of northeast India in Guwahati on 10 September 2010; on the issue of out-migration from Assam see Muktiar and Sharma (2019)
22 Vagholikar and Das (2010: 4, 8).
23 Ibid.: 3.
24 Sharma (2008: 4).
25 Sharma (2013).
26 Vagholikar and Das (2010: 7).
27 Ibid.
28 Under a system known as 'Inner Line' introduced during the colonial period, every outsider needs a permit to enter Arunachal Pradesh. Besides, outsiders are not allowed to purchase and become permanent owners of land in AP.
29 Ibid.
30 Goswami (2015: 1).
31 Baruah (2012: 4).
32 'GoM decides to scrap NTPC hydel project on Bhagirathi river' (http://articles.economictimes.indiatimes.com/2010-08-20/news/27599592_1_hydel-project-bhagirathi-river-loharinag-pala); Tripathi, 'Politics of ecology' (www.frontline.in/environment/conservation/politics-of-ecology/article4756368ece).
33 Tiwari (2010: 1, 4).
34 'Pratibedonar Stithit Atal Asomar Bisheshagna Samiti' (The Assam expert committee sticks to its report), 2012.
35 Chakravartty (2013).
36 'Akwé: Kon guidelines' (www.cbd.int/doc/publications/akwe-brochure-en.pdf).
37 Pertin (2013: 134–138).
38 'Bandha Hobo Lagiba Namoni Siyang Brihat Nadibandh' (The Lower Siyang big dam must be stopped), 2010.
39 'Arunachalat Hinsatmak Rup Loise Nadibandh Birodhi Andolone . . .' (Anti-dam movement assumes violent form in Arunachal), 2012.
40 'Arunachalat Sakalo Nadibandhor Rajahua Sunani Batil' (The public hearing of all dams in Arunachal cancelled), 2012.

41 Mimi (2018: 218–230).
42 Mazoomdar, '6 years, 2 rejections later, India's largest hydro project cleared' (http://indianexpress.com/article/india/india-others/6-years-2-rejections-later-indias-largest-hydro-project-cleared/).
43 'With anti-dam mantra on lips, Tawang monks hit the streets' (http://timesofindia.indiatimes.com/city/guwahati/With-anti-dam-mantra-on-lips-Tawang-monks-hit-the-streets/articleshow/18842465.cms); 'Tawang threatened by 15 proposed dams' (www.huffingtonpost.com/urmi-bhattacharjee/tawang-threatened-by-15_b_2502513.html?ir=India&adsSiteOverride=in).
44 'Tawang hydel project gets MoEF approval' (http://archive.indianexpress.com/news/tawang-hydel-project-gets-moef-approval/1069846/).
45 Ibid.
46 Mazoomdar, '6 years, 2 rejections later, India's largest hydro project cleared' (http://indianexpress.com/article/india/india-others/6-years-2-rejections-later-indias-largest-hydro-project-cleared/).
47 Dodum (2014: 17).
48 Goswami (2014: 15)
49 Chakraborty (2015).
50 Mazoomdar, '6 years, 2 rejections later, India's largest hydro project cleared' (http://indianexpress.com/article/india/india-others/6-years-2-rejections-later-indias-largest-hydro-project-cleared/).
51 Shrivastava, 'Environment ministry clear controversial Demwe hydroelectric project' (www.downtoearth.org.in/news/environment-ministry-to-clear-controversial-demwe-hydroelectric-project).
52 Ibid.
53 'New Assam government vows to oppose big dams' (http://thewire.in/2016/06/03/new-assam-government-vows-to-oppose-big-dams-40291/).
54 'BJP calls for halt to work on Subansiri dam' (www.assamtribune.com/scripts/detailsnew.asp?id=nov1810/at07).
55 'No clearance for Arunachal hydel projects' (www.telegraphindia.com/1150702/jsp/northeast/story_29019.jsp#.Vb9BSyLoRtE).
56 Ibid.
57 Rina (2015: 1)
58 Dodum (2015: 1).
59 'Plea on Dibang dam nod' (www.telegraphindia.com/1150803/jsp/northeast/story_35133.jsp#.WB3mvdJ97IU).
60 Kukreti (2017).
61 Karmakar, 'Arunachal mega dam near pilgrimage cleared, environmentalists see red' (www.thehindu.com/news/national/other-states/arunachal-mega-dam-near-pilgrimage-cleared-environmentalists-see-red/article25066578.ece).
62 Konwar (2017: 15).
63 'APW recommends seismology expert for lower Subansiri project' (https://timesofindia.indiatimes.com/city/guwahati/apw-recommends-seismology-expert-for-lower-subansiri-project/articleshow/64091545.cms).
64 Assam NGO moves SC, seeks stay on NGT order on power project, https://www.newindianexpress.com/nation/2019/nov/15/assam-ngo-moves-sc-seeks-stay-on-ngt-order-on-power-project-2062201.html
65 Sharma (2015: 22, 39–40).
66 Ibid.
67 Baruah, 'Dam wrong' (www.hindustantimes.com/india/dam-wrong/storyERxm-9KrOEBU5UNna8geZ1O.html).
68 For delails on this see Sharma and Borgohain (2019: 17–20)

69 Ibid.
70 Dai (2010).

References

Ahmed, F. (2003) 'Violations and evasion: The public hearing at Gerukamukh.' *The Ecologist Asia*, 11 (1), p. 30.

'APW recommends seismology expert for lower Subansiri project.' (2018) Retrieved from https://timesofindia.indiatimes.com/city/guwahati/apw-recommends-seismology-expert-for-lower-subansiri-project/articleshow/64091545.cms

'Arunachalat Hinsatmak Rup Loise Nadibandh Birodhi Andolone' (Anti-dam movement assumes violent form in Arunachal). (2012, April 19) *Amar Asom*, Guwahati, pp. 1, 8.

'Arunachalat Sakalo Nadibandhor Rajahua Sunani Batil' (The public hearing of all dams in Arunachal cancelled). (2012, April 19) *Amar Asom*, Guwahati, p. 3.

'Bandha Hobo Lagibo Namoni Siyang Brihat Nadibandh' (The Lower Siyang big dam must be stopped). (2010, December 20) *Asomiya Pratidin*, Guwahati, pp. 1, 8.

Baruah, S.K. (2005) *Durable Disorder: Understanding the Politics of Northeast India*. Oxford University Press, New Delhi.

Baruah, S.K. (2010, September 26) 'Dam wrong.' Retrieved from www.hindustantimes.com/india/dam-wrong/storyERxm9KrOEBU5UNna8geZ1O.html

Baruah, S. (2012, January 22) 'Lower Subansiri and the politics of expertise.' *The Assam Tribune*, Guwahati, p. 4.

Baruah, S.K. (2012, March 14) 'Arunachal had warned NHPC on dam project.' Retrieved from www.hindustantimes.com/newdelhi/arunachal-had-warned-nhpc-on-dam-project/article1-824975.aspx

Baviskar, A. (2004) *In the Belly of the River: Tribal Conflicts Over Development in the Narmada Valley*. Oxford University Press, New Delhi.

'BJP calls for halt to work on Subansiri dam.' (2010, November 18) Retrieved from www.assamtribune.com/scripts/detailsnew.asp?id=nov1810/at07

Chakraborty, S. (2015, June 25) 'Rejected by experts, Dibang hydro project gets green nod.' *Business Standard*. Retrieved from www.business-standard.com/article/economy-policy/rejected-by-experts-dibang-hydro-project-gets-green-nod-115062400037_1.html

Chakravartty, A. (2013, March 12) 'Subansiri dam unsafe: Experts' committee.' Retrieved from www.downtoearth.org.in/news/subansiri-dam-unsafe-experts-committee-40558

Dai, M. (2010, September 26) 'This development is like an invasion.' Retrieved from www.hindustantimes.com/india/this-development-is-like-an-invasion/story-4xK57DVSglCrTyTrhHbZ2I.html

'Dam-induced flood & nature's fury devastate North Lakhimpur.' (2017, July 11) *The Assam Tribune*. Retrieved from www.assamtribune.com/scripts/detailsnew.asp?id=jul1117/state050

Dodum, R. (2014, November 1) 'Protests over Dibang dam fall silent.' *The Telegraph*, Guwahati, p. 17.

Dodum, R. (2015, July 2) 'No clearance for Arunachal hydel projects.' *The Telegraph*, Guwahati, p. 1.

Goswami, D.C. and Das, P.J. (2003) 'The Brahmaputra river, India.' *The Ecologist Asia*, 11 (1), pp. 15–18.

Goswami, R. (2014, October 21) 'NE voice muffled to clear Dibang path.' *The Telegraph*, Guwahati., p. 15.

Goswami, R. (2015, June 1) 'Dibang roadblocks remain.' *The Telegraph*, Guwahati, p. 1.

Guyot-Réchard, B. (2013) 'Nation-building or state-making? India's North-East Frontier and the ambiguities of Nehruvian developmentalism, 1950–1959.' *Contemporary South Asia*, 21 (1), pp. 22–37.

Karmakar, R. (2018) 'Arunachal mega dam near pilgrimage cleared environmentalists see red.' Retrieved from www.thehindu.com/news/national/other-states/arunachal-mega-dam-near-pilgrimage-cleared-environmentalists-see-red/article25066578.ece

Konwar, R. (2017, October 17) 'Panel to study hydel project.' *The Telegraph*, Guwahati, p. 15.

Kukreti, I. (2017) 'NGT suspends forest clearance for Arunachal hydropower project.' Retrieved from www.downtoearth.org.in/news/ngt-suspends-forest-clearance-for-arunachal-hydroelectric-project-58963

Mazoomdar, Jay. (2014) '6 years, 2 rejections later, India's largest hydro project cleared.' Retrieved from http://indianexpress.com/article/india/india-others/6-years-2-rejections-later-indias-largest-hydro-project-cleared/

Mimi, R. (2018) 'Dibang multipurpose project: Resistance of the Idu Mishmi.' In K.J. Joy, P.J. Das, G. Chakraborty, C. Mahanta, S. Paranjape and S. Vispute (Eds.), *Water Conflicts in Northeast India*. Routledge, London and New York, pp. 218–230.

Muktiar, P. and Sharma, C.K. (2019) 'In search of a better future: Nepali rural out-migration from Assam.' *Sociological Bulletin*, 68 (3), pp. 307–324.

Pertin, Azing. (2013) 'Lower Siyang hydropower project: A peaceful valley erupts.' In P.J. Das, C. Mahanta, K. Joy, S. Paranjape and S. Vispute (Eds.), *Water Conflicts in Northeast India: A Compendium of Case Studies*. Forum for Policy Dialogue on Water Conflicts in India, Pune, pp. 134–138.

'Plea on Dibang dam nod: NGT admits duo's petition challenging clearance.' (2015, August 3) *The Telegraph*. Retrieved from www.telegraphindia.com/1150803/jsp/northeast/story_35133.jsp#.WB3mvdJ97IU

'Pratibedonar stithit atal asomar bisheshagna samiti' (The Assam expert committee sticks to its report). (2012, January 17) *Amar Asom*, pp. 1, 8.

Ramesh, J. (2005) 'Northeast India in a new Asia.' *Seminar*, (550), pp. 17–71.

Rina, T. (2015, September 15) 'MOEFCC on Massive clearance spree of Arunachal hydropower projects.' *The Arunachal Times*. Itanagar, p. 1.

The Report of the World Commission on Dams. (2000) *Dams and Development: A New Framework for Decision-Making*. Earthscan Publications Ltd, London and Sterling, VA.

Sharma, C.K. (2008, September 3) 'Brihat Nadibandh Prasanga: "Bishesh" aru Sadharan Gnan' (On big dams: 'Specialised' knowledge and common sense). *Dainik Janambhumi*, Guwahati, p. 4.

Sharma, C.K. (2013) 'Krishak Mukti Sangram Samity and its struggle: The new peasant assertion in Assam.' In A. Baruah and S. Sengupta (Eds.), *Social Forces and Politics in North East India*. DVS Publishers, Guwahati, New Delhi, pp. 130–176.

Sharma, C.K. (2015) 'Damming the Brahmaputra – Whom to blame?' *One India One People*, 18 (7), pp. 22, 39–40.

Sharma, C.K. (2017) *Political Economy of the Conflicts Along the Assam-Arunachal Pradesh Foothill Border*. ActionAid India, Guwahati.

Sharma, C.K. and Borgohain, B. (2019) 'The New Land Settlement Act in Arunachal Pradesh: Road to empowerment or dispossession?' *Economic and Political Weekly*, 54 (23), pp. 17–20.

Shrivastava, K.S. (2015) 'Environment ministry clear controversial demwe hydroelectric project.' Retrieved from www.downtoearth.org.in/news/environment-ministry-to-clear-controversial-demwe-hydroelectric-project

'Tawang hydel project gets MoEF approval.' (2013, February 6) *The Indian Express*. Retrieved from http://archive.indianexpress.com/news/tawang-hydel-project-gets-moef-approval/1069846/

Tiwari, R. (2010, October 8) 'In note to PM, Jairam takes on govt, puts question marks on NE projects.' *The Indian Express*, New Delhi, pp. 1, 4.

Tripathi, P.S. (2013, June 14) 'Politics of ecology.' *The Frontline*. Retrieved from www.frontline.in/environment/conservation/politics-of-ecology/article4756368.ece

Vagholikar, N. and Das, P.J. (2010) *Damming Northeast India*. Kalpavriksh and Aaranyak, Pune and Action Aid India, Guwahati.

'With anti-dam mantra on lips, Tawang monks hit the streets.' (2014, March 7). *The Times of India*. Retrieved from http://timesofindia.indiatimes.com/city/guwahati/With-anti-dam-mantra-on-lips-Tawang-monks-hit-the-streets/articleshow/18842465.cms

Part II

COMMUNITY, POLITICS AND LIVELIHOODS

5

NILGIRI BIOSPHERE RESERVE

Reflections from the field

Ritambhara Hebbar[1]

Most discussions concerning the environment are highly technical and lack any reference to people who are directly implicated in the proposed plans and projects. In this regard, the Nilgiri Biosphere Reserve (NBR) is no exception. There is hardly any reference to the rich cultural diversity of the region and its relevance for conservation agendas within NBR. Nilgiris is by far, as Hockings puts it, 'the most intensively studied part of rural Asia east of the Holy Land' with over 3,000 books and articles, that is, more than 'three publications per square mile' of the region (1989: vi). How are we to then comprehend the complete neglect of this tremendously rich literature on the biogeography of the Nilgiris in the environmental deliberations on the region? Could this be read as a part of a larger politics of assimilation that dominate the tribal question in the country?

The chapter explores this paradox to argue how environmental projects have significantly reproduced the larger politics against tribal communities in India. The chapter has two sections. Based on a reading of official documents pertaining to the creation and the action plan of the NBR, section one seeks to unravel its politics. The politics of conservation emanates from the central assumption that accompany most conservation efforts, that managerial fixes and technical know-how are of a higher order, workable and reproducible across sites. The case of the Nilgiris is a case in point. There is no discussion on the local history and political economy, the close association of the local tribal communities with the forests and their significant role in sustaining biodiversity conservation in the Nilgiris in the official documents of the NBR. Nor is there any enunciation of a scientific vision, let alone a political vision, behind the formation of biosphere reserves (BRs). In fact, I would argue that politics of conservation is precisely this silence over difficult political questions on how conservation agendas would contend with various actors with competing interests in controlling and managing natural resources. This lack of history and politics is also reflected in the directionless course that conservation efforts have taken in NBR. The second section draws on my research on the Nilgiris. I present excerpts from interviews with tribal and local activists that elucidate the government practices associated with the reserve. The focus is on the nature of governance in the reserve and the various ways by which the local tribal communities have been marginalised within the reserve and denied any claims over the forest. The section presents environmental and conservation projects as modes of governance that introduce newer institutional processes and practices in the control and management of natural resources. New government practices now frame the discussion on conservation of biodiversity in the Nilgiri.

Nilgiri Biosphere Reserve: a case of green politics

There are about 14 BRs in India. Nilgiris was the first BR to be set up in 1986 and recognised as one of the biodiversity hotspots in the world with a wide variety of species, both plant and animal, particularly large mammals such as tigers, elephants and monkeys. Spread over the three southern states of Tamil Nadu, Karnataka and Kerala, the NBR comprises of different environmental projects initiated with a specific purpose. It includes

> Rajiv Gandhi National Park (Nagarahole) and Bandipur (Karnataka), Wynaad, the slopes of Nilambur, Silent Valley and Siruvani Hills (Kerala), and Dr. J Jayalalitha Wildlife Sanctuary (Mudumalai), Nilgiris and Mukurthi [Sic.] (Tamil Nadu). Of these Bandipur, Nagarhole and Silent Valley are National Parks, Mudumalai, Wynaad and Mukurthi [sic.], are Wildlife Sanctuaries. Bandipur is a Tiger Reserve, and also the single largest Protected Area within the NBR.
>
> (Daniels 1996: 5)

Globally, deliberations on establishing biospheres started in the early 1970s under the auspices of UNESCO, and as a part of the International Coordinating Council of Man and the Biosphere Programme (MAB). At first glance, the concept of BRs as enunciated by the council appears imprecise in terms of its larger purpose. To quote from the report of the Task Force on Criteria and Guidelines for the Choice and Establishment of Biosphere Reserves (1974: 6):

> Biosphere reserves are not meant to be substitutes for national parks and equivalent reserves. Biosphere reserves can coincide with or incorporate national parks or equivalent reserves but they could also include areas which do not conform to the IUCN[2] definition of National Parks. They may also include buffer zone areas where manipulative research may take place. The most significant and distinct characteristic of biosphere reserves, however, will be their

> links by international understanding on purpose, standards and exchange of information and personnel. . . . It is recognised that the success of the program will depend upon public information, education and training of personnel within all countries concerned. Full use should be made of the educational resources of existing sites, especially the national parks and reserves, and model programmes should be developed within these sites in co-operation with local educational systems.

The larger aim of forming BRs was of conserving genetic diversity of species and evolving model programmes through systematic scientific and environmental research in collaboration with local education and research institutions in the region. The BRs were a means to bring about a commonality of scientific purpose and vision across the world regarding conservation. It was about building an intellectual corpus and consensus through a worldwide network of intergovernmental and non-governmental organisations as per the guidelines laid out by the council. The BRs thus could well be seen as a research-oriented initiative that tried to bring about an integration of thought and purpose across the world on the question of environmental conservation.

The reserve was organised into three zones: core, buffer and man-modified landscapes, which is also known as the manipulation (forestry and agriculture) zone. The core zone is the secluded area, with no or minimum human activity. The buffer zone is the area that acts as a barrier between the core and the manipulation (forestry and agriculture) zone, and also the site of scientific research and experimentation. The manipulation zone, which was further divided in the case of Nilgiris, is the tract left accessible for development activities. The committee constituted by the Department of Environment, government of India to demarcate the various zones of the Nilgiri Biosphere in 1984 identified, other than the core and buffer zone, six more zones: Manipulation (Forestry) Zone; Manipulation (Tourism) Zone; Administrative Zone; Manipulation (Agriculture) Zone; Restoration Zone; and Link Zone (Department of Environment 1984). These zones included activities such as forestry-related programmes, ecotourism, infrastructure for development of administrative facilities, agriculture programmes to promote cultivation of genetic diverse crops and plants and afforestation programmes both to restore degraded forests in the region and to maintain continuity of natural vegetation through the reserve.

In December 1986, a proposal for an action plan was prepared by the Indian Institute of Science (IIS) for a scientific programme in the Nilgiri Reserve. It laid out the details of the network of research institutions,[3] laboratories, universities and government and non-government organisations which would be involved in the programme, particularly in the various surveys of the region, and the related expenses of setting up administrative facilities (Centre for Ecological Sciences 1986). The initial interest was of creating a data bank on the Nilgiris, on its history, culture, economy, administration and natural environment. Research on tribal communities formed a significant part of gathering 'ethnobiological' information, on the 'micro cultural diversity' as evidenced among the local tribes in the region. The idea was to get a 'bird's eye view' on the Nilgiris as against the worm's eye view, that is, collect 'ethnobiological information' without protecting the cultural contexts within which such information is located (Gadgil et al. 1986: 11–12). The involvement of voluntary organisations such as the Paryavaran Vigyan Kendras to promote eco-development and spread awareness about the 'scientific basis of the BR programme into the commonly understood format such that it is assimilated by the people of the area and they become partners in the process' was a clear indication of how local tribal communities were sought to be tutored into thinking scientifically about the environment (ibid.: 22). Subsequently, in 1994, a survey was conducted by IIS on BRs in the country, which expressed disappointment with the lack of coordination between the three states, forest departments and between the government institutions and the local communities in the case of Nilgiris. According to this survey, BRs had failed to 'reconcile development activities and conservation, to harness the knowledge of local communities, to involve such communities in management, or in other ways to achieve the objectives originally set for BRs by UNESCO' (Kothari et al. 1995: 2762). But the critical question is whether the original objectives were these at all.

In 1988, the World Conservation Monitoring Centre in collaboration with the IUCN's Commission on Natural Parks and Protected Areas prepared directories detailing individual protected areas in the country (World Conservation Monitoring Unit 1988). The NBR was a part of this exercise. The draft proposal prepared on the Nilgiris describes, among other topographical specificities, the nature of vegetation, conservation management and the management problems within each of the wildlife sanctuaries and national parks located within the NBR.

The proposal reveals how the objectives of BRs were always against local forest communities and their presence in the forests. For instance, listing the specificities of the various parks and forest reserves within the NBR, the draft made a case for conservation management in the region by moving tribes from within the core and buffer areas of the reserves. About 20 tribes were listed as inhabiting the proposed reserve. These were the Adujans, Alars, Arandans, Cholanaiks, Irula, Kalanaiks, Kunduradigans, Kurchiyans, Kurumans, Kurumba, Malmuthans, Mullu Kurumbas, Pahinaikas, Paniyans, Pariyans, Tackanadmuppans, Todas, Uralikurulians and Wayanad Kaders.

Tribes were systematically targeted within the environmental discourse on conservation. Conservation management plans were set against what were presented as main threats of cultivation and grazing, largely practised by local tribes, and by fires, conflicting interests of forestry operations and water resource development. For instance, in the case of Bandipur, the Kabini reservoir was not construed as a hindrance to conservation, while the cattle grazing around it was; wildlife tourism was permitted within conservation management, while pastures for livestock of local tribes were not. There was then a systematic alienation of tribes from within the reserve, as new government practices were instituted to control and manage natural resources. Even where state and private development activities such as hydroelectric projects (as in the case of Silent Valley), and commercial plantations (specifically Wayanad), emerged as critical conservation concerns, the plan appeared to be ineffective in stalling such activities. The plan privileged a set of development and commercial practices patronised by the state through a managerial approach. It supported the politics that dominated the region and was responsible for the destruction of its biodiversity. The model proposed by the BR has significantly shaped the politics on the question of resource management, and is also shaped by the politics that dominates the region.

There was no discussion over the possible challenges in the implementation of such a programme in a region dominated by plantations. Interestingly, the timing of the declaration of the reserve more or less coincided with the publication of the Ministry of Labour and Rehabilitation, Government of India report in 1984 on the working and living conditions of plantation labour in the Nilgiri region spread over the three states of Tamil Nadu, Karnataka and Kerala. The study focused on how tea, coffee and rubber plantations in the three states covered 60 per cent of the total area and 50 per cent of the labour employed therein (Raman 1986). A major section of the plantation labour is from tribal communities. For instance, in the Nilgiri district, over 71 per cent of the Irula work in tea and coffee plantations; about 19 per cent of Kattunayans work in tea and spice plantations; and about 69 per cent of Paniyans work in plantations and private agricultural fields. They are engaged in collection and selling of firewood from non-reserved forest areas to government-run cooperative depots in Gudalur and Pandalur towns of Nilgiri district (Parthasarathy 2007). The proliferation of plantations attracted migrants and plantation workers, and as a consequence increased the population of the region (Venugopal 2004: 76). The extension of monoculture plantations, such as tea, coffee, cinchona, eucalyptus and wattle, destroyed the moist evergreen forests, shola forests and grassland. Nilgiris has become the largest tea-growing region in South India with 60,000 small tea growers in the region, and it accounts for one-tenth of the national tea production (ibid.: 79). Eucalyptus is another such commercial crop which has brought about water scarcity in the region. It was introduced in the forests of Nilgiris in the 1850s by the British to use as fuelwood to protect native varieties from being destroyed. But in 1950s, the indigenous forests were being destroyed to plant eucalyptus to meet growing demand from the synthetic rayon factory. Apparently, the damming of rivers in Nilgiris produces 1000 MW of power, which accounts for 40 per cent of the hydroelectricity in the state.(Ibid.: 76).

There has been various development related illegal activities within the BR. There have been regular encroachments of forest and revenue land for tea plantations, with the government also sponsoring one of the plantations to provide employment to Sri Lankan Tamil repatriates who were settled in the region in the 1970s. There have also been international projects such as the Indo-German project of potato cultivation in the steep slopes of the Nilgiris which have led to landslides and siltation of water bodies. Roads have been laid across the NBR that cut across the sanctuaries. For instance, the road from Coimbatore/Udhagamandalam and Mysore–Bangalore cuts through Mudumalai and Bandipur. There is also organised poaching of sambar (cervus unicolor) and chital (axis) deer in the Nilgiris (Daniels 1996).

Nilgiris has been a popular tourist destination since the 1980s. The conflict in Kashmir redirected tourists to the Nilgiris, with their numbers increasing to more than a million in the year 2000 (Venugopal 2004: 74). The number of tourists arriving at Nilgiris per year since 2000 has expanded more than a million (ibid.: 74). There are more than 500 lodging and boarding facilities including hotels, lodges, travel and entertainment-related services in Nilgiris to cater to the growing population of tourists in the region which goes up to 0.3 million in the month of May alone (ibid.: 76). There have been frequent landslides in the Nilgiris (in 1993, 1995 and 2002), and the rise in pollution of subsurface water due to an inappropriate drainage system has weakened the stability of the slopes (ibid.). Clearly, indiscriminate development activities have been allowed since 1986 in the Nilgiris that have compromised on the principle reason for which the NBR was created – safeguarding the biodiversity of the Nilgiris.

Seeing red – the tribal situation in the Nilgiri

My fieldwork in the Nilgiris, and the reference here is to Masinagudi, Pandalur, Gudalur, Sathyamangalam (Erode), Ooty, B. R. Hills and Wayanad, revealed a complex web of issues that implicated the lives of the local tribal communities. For one, there was a history to their marginalisation dating back to British rule, the coming of the plantations and the commercialisation of tribal dominated regions. There were concerns over the lackadaisical attitude of the government towards them, illustrated in the arbitrary system of classification of tribes with many sections of the same tribe in another district denied the scheduled tribe status, and in the complete disregard of an ongoing demand among tribal organisations to consider tribal-dominated areas in the region within the fifth schedule. There were also repeated references to the growing presence and involvement of NGOs in the implementation of programmes and projects, and the consequent lack of accountability and responsibility within the state government institutions. Among all these issues were concerns over their displacement from within the core and buffer zones of sanctuaries, tiger and elephant reserves, the fate of their patta[4] lands within these zones, their evoking of the Forest Rights Act (FRA)[5] to stake claims on their rights on land and the role of the forest department in thwarting their efforts to do so. Clearly, the environmental agenda, through its various government strategies, undermined tribal protest and the core issues that tribes are facing in the region. Conservation is a site of politics and there is a need to demystify conservation efforts and accompanying scientific discussions to re-examine them as a mode of governance. As a set of government practices, conservation initiatives reveal the pursuance of the very opposite of its ostensible purposes.

Presented next are some excerpts from interviews that illustrate how environment as a subject is entangled in a web of government practices and steeped in local politics. They expose the myopic vision encased in environmental agendas. I have divided my interviews thematically to reveal how conservation agendas appear to communities living in the reserve, as modes of governance that have introduced institutions and processes to deny their claims on land and forests.

Politics of conservation

The following are excerpts from interviews with tribal and local activists from Gudalur, Pandalur and Mudumalai on the politics of conservation, the role of international environmental NGOs in the region and their involvement with the state forest and revenue department in undermining tribal rights in the forest as well as projecting tribes as harmful to the well-being of the forest. The excerpts also reveal the extent of commercialisation and politicisation within the reserve.

> *[Gudalur]*
>
> The three communities – tribes, Malayalis, and the repatriates – constitute the majority population of Gudalur, Pandalur and Masinagudi. There are 4 lakhs [400,000] people all together. The repatriates were the largest community, constituting more than 2 lakhs [200,000], with about 80 to 85,000 Malayalis. The tribal population is 20 to 25,000. The government's figure for tribes in Nilgiris is 28,000. Apart from these three communities, there are Kannadigas, Chettis and some mixed population.
>
> The land is primarily in the hands of the big estates; then there are about 300 small estates which control 2,000 to 4,000 acres. The smaller estates are predominantly owned by Malayalis. The tribals

and the dalits have roughly an acre (each). Otherwise mostly they have 5 to 10 cents (1 cent equals 440 square feet). And the tribes hold land that has been traditionally theirs. The government has not given them the patta. The dalits do not have patta for their land either. Few tribes (mostly Mullu Kurumba) have patta, that they got during the British rule. . . . Land had been leased to 12 companies for coffee cultivation. Since coffee didn't grow well here, they replaced coffee with tea. The land held by the companies is 32,000 acres. And all of it is evergreen forests. The Tamil Nadu government, using section 17, Land Ceiling Act, attempted to take back land from the companies; the companies refused to hand over the land stating they had leasing rights. They went to court and asked the government to either give them patta or provide compensation. The Court said that the land that was under their control could be used by them until the case got over, but that they shouldn't encroach on more land. The companies ignored this order and gradually they have taken over a lot of land. The forest department is with them.

There are some very essential varieties of trees in Gudalur forests. On an average about 32 lorries loaded with these trees are taken out through Kerala. They are exported out of the country. None of the environmental groups here take up the issue of illegal felling of trees by companies.

Conservationists claim that these settlers are destroying the forests and that that the tribes are killing the animals by hunting. They keep labelling the people as encroachers. The first encroachers were the people of Nilambur, then the British and now it is the 12 companies. But the companies are never questioned. The forest department keeps harassing the people living here now; they come and break down their houses, file false cases on them, arrest people and damage crops.

In addition to the Forest Department, we now have to deal with the conservationists. They claim that it is the people who are destroying the forest. This makes building a case to evict the people easier for the government. Our agitation is to clarify as to who is truly destroying the forest. We collect data to substantiate our statements. We keep track of how many lorries of trees are being taken out (from check post logs), who are the land grabbers, how much is an estate legally allowed to own. . . . For instance, in Gudalur there are four timber sawmills located within the forest. As per the Supreme Court directions this is a violation; the order states that you cannot have any construction within 1 kilometre of a National Forest. The conservationists have links at high levels – with the superintendent of police, collector, and inspector general.

This excerpt illustrates how the conservation agenda (or the Man and the Biosphere Programme) is caught in a web of litigation, which also reveals the politics of natural resource management within the reserve. There are host of actors, which include estate and plantation owners, state governments, the judiciary, local communities, lawyers, dominant NGOs, the forest department and migrant populations that seek to retain hold over the land and forest resources in the area. The Act referred to in the excerpt is the Janmom Act 1969–1974. Section 17 of the Act empowers the government to renew or terminate leases given as independent holdings (Janmom land). This has been used by local officials to facilitate land grabs in the region. Conservationists and big environmental organisations have worked in collusion with large estate owners and forest officials, allowing illegal occupation of land in the Reserve. However, the same clause has been used for evicting local tribes from their land.

> *[Mudumalai]*
>
> There is a Singara Reserve Forest and then the Mudumalai Wildlife Sanctuary. In 1998, they proposed to merge these two (in the context of the elephant corridor). In 1991 they wanted to declare the entire land as private forest. Once that happens, you will have to seek their permission for any activity including planting a tree. An estate owner here has 1,500 acres. For about 300 acres, there is coffee plantation and the rest are lying fallow. The estates need to be declared as private forest and you need to have regulations so that the forest is not destroyed. I have 2 acres of land where I am cultivating.
>
> This cannot be considered as private forest. They are declaring every 1 cent or 2 cents of land as private forest. Even the road has been declared as private forest. There is a government order (GO). The police station and church have been made private forest. We fought back in 1991 against this but did not file a case. But now all of our land has been declared as private forest. We cannot buy or sell the land. We cannot build on our land. They say that even cultivation is not allowed.
>
> In 1998 we filed a case against the merger of Singara Reserve Forest and the Mudumalai Wildlife Sanctuary. If this comes through, we will not have any rights, as we cannot even enter inside a wildlife sanctuary. There are people living inside the Mudumalai.

The local tribal population find themselves sandwiched between the core zone and private forests, unable to access the forest on either side. The Tamil Nadu Preservation of Private Forest Act 1949 is one of the most critical regulations through which the state government has retained its control over land in the region. Any occupation and sale of land that falls within the provision of private forests require prior permission and approval. Unless ratified under the Private Forest Act by the government, land holdings within private forests are illegal. By declaring cultivable and occupied land as private forests through a recent government of India notification, the state government has assumed the power to evict and reclaim the land within the area. Lack of clarity about government notifications that are revised intermittently have added to the anxiety among local tribal and dalit communities. With the declaration of the elephant corridor and tiger reserve, the locals feel further trapped. They were not informed of these changes or asked to vacate the area. They continue to live in the area fearing impending eviction. They have been systematically excluded both in the vision and practice of conservation. The FRA of 2006 has given them some hope of claiming rights on land within the zones and reserve. The local hamlets have organised themselves into a forest rights committee, which the forest department does not recognise. The forest department prefers to work through the big NGOs that are involved in development work in the area.

Forest Rights Act

[Gudalur]

When the Forest Rights Act came, we started placing boards everywhere. Now the Forest Department is more scared. Previously, if any official came and questioned as to why we were cultivating here, the people used to be speechless. They used to quote the Supreme Court order, to stop us from cultivating or even plucking grass from the forest. Now, we can boldly claim that this section of the FRA allows us to cultivate. The forest is ours. Our arguments are stronger. They are unable to do anything. The FRA is protection to us. If not for this, they jail us and call us terrorists. Now they are unable to do that. The government places a certain importance on the Forest Rights Act; even if they don't implement it, they are scared of it. Even in the case of the tiger reserve and the elephant corridor we have used the FRA. The Act says that without the gram sabha resolution you cannot take any project forward. The government has not followed this procedure. We hold the government responsible for such lapses.

The people come together, about 500 to 1,000, against such illegalities. It was more difficult to organise the people before 2006. But now the people and also the political parties around here are convinced. In December 2009, 1 lakh [100,000] people gathered for a demonstration – dalit repatriates, adivasis, Malayalis and others participated.

> We have formed the gram sabhas for every settlement where there are dalits and adivasis. . . . Like in the 5th and 6th schedule, we have established gram sabhas for every settlement. The people of the settlement will decide. At the moment, the government is accepting claims only from the tribes. The forest dweller's claims are not being accepted because they do not fall within the purview of forest dwellers.
>
> The Chettis are the only forest dwellers who qualify as per the Act. In Tamil Nadu, the implementation of the Forest Rights Act is poor. They state that so far, they have got 250 claims from Nilgiris, whereas we have filed more than 5,000 claims. They neither process claims nor respond to us.
>
> There is one case filed by the Forest Department and NGOs at the Madras High Court fighting against implement the FRA. In Thirunelveli district, a Zamindar who owns thousands of acres of land also has filed a case at the Madurai High Court against the implementation of the FRA. In Pune and six to seven other places across the country, similar cases against the implementation of FRA have been filed. The NGOs are behind many of the cases. In Madras High Court, we have challenged the case. The case is going on.

Interestingly, the resistance to Forest Rights Act has been mostly from the forest department, local zamindars (large landowners) and NGOs. This exposes the complex contestations over land and forest in the area. It also reveals the changing contours of governance, where actors other than state departments now actively participate in determining the course of conservation practice. Conservation is the effect of this complex web of litigation and local politics rather than an envisioned programme of action.

While the local tribal communities and adivasis have mobilised around FRA to claim titles for their land, they face the wrath of conservationists. The next excerpt from an interview of an activist in Masinagudi, Tamil Nadu, illustrates this. The main tribal communities in Masinagudi are the Irula, Kattunaicka, Paniya and Kurumba. There is a sizeable population of dalits, adi kannadigas and adi dravidars. Some of the Irula, who settled in the area earlier, and Kurumba have land, but now with divisions over generations, have barely a half acre per person. Most of the estates are owned by Marwadis (Ethnic category comprising of trading communities, primarily from Rajasthan), Malayalis and Muslims.

> *[Masinagudi]*
>
> In February 2009, the Forest Department started talking about the Forest Rights Act. This area has been included within the elephant corridor. It lies within the core zone. We are surrounded on three sides by land that the Forest Departments wants to declare, and on

one side by vast private property. Our land for cattle grazing and firewood collection is entirely dependent on the core zone. The Forest Department is saying that not only can we not use the land but we cannot even walk through it. How can we survive? They fine us for going through the forest. I have paid a fine myself. Shouldn't they discuss with us before declaring land that our livelihood is dependent on as a core zone? We still take our cattle there for grazing. The government should have checked with us about it. We do not have a choice. They do not have any idea or understanding when they bring out a rule. They have decided that a certain area is the core zone for the tiger reserve, areas including our villages are buffer zones, and we are expected to act accordingly.

Well, when they were legally declaring the core zone, they should have asked us if we could find alternatives or we could move anywhere else. None of our rights have been settled so far. We call it the 'illegal' tiger reserve. We will only accept the tiger reserve when our rights are settled.

There are about 10 to 15 hamlets here. When we go for discussions, either all villagers assemble or representatives from each hamlet assemble and go together. There are women's groups, drivers' union and adivasi vana urumai iyyakkam (forest rights committee).

The forest rights committees have been formed by the villagers themselves here, but they are not recognised by the government. The collector refused to recognise the committees that the villagers formed. The collector called for a meeting with the NGOs and demarcated area for each NGO to work in and form the forest rights committee. In the beginning, we thought that the NGOs had a decent relationship with the state and they could mediate between the people and the state. They use to collect all the information from us and then one day filed a court case against us using our information. . . . This is unfair. If they wanted, they could have taken a stand in the open, or could have come for discussions with us. At this point, the department had preliminary papers to evict us. The police came here, and pulled out the fences. That is when we filed a case in the Madras High Court. We demanded the implementation of the FRA in Masinagudi. The case is going on.

The rights include provision of patta for individual owners and then community rights for grazing. Our village land is in the buffer zone so we can continue to live here. But there is no livelihood, no grazing land and firewood and other forest produce collection. And we don't have mobility. They moved the entire issue on a technical ground. On 31 December 2006, the Forest Rights Act was brought about. On 28 December, the tiger reserve was announced. Hence,

> for implementing the tiger project, they state they don't have to settle rights as per the FRA. It is their technical advantage.

These excerpts illustrate how the local tribal communities were hoping to use the FRA to stake claims on the land they have been cultivating and avoid eviction from within the forest. However, environmental rules and regulations related to the tiger project and the elephant corridor have been cited to push FRA claims of local tribal communities to the back burner. The NGOs have played a critical role in the ensuing politics over FRA, in underestimating the number of FRA claimants, as well as in not recognising individual and collective rights within the forest.

Dominance of NGOs

> *[Gudalur]*
>
> There are many NGOs here. . . . Their work is solely based on projects and funds. If there is any fund for hair, they will develop a project around it. That is how lame they are. They are completely useless and are also a huge problem. They divert the issues away from the focal point. NGO 'A' does awareness work and welfare programs. They hold rallies with a banner asking for implementation of Forest Rights Act, take photos and have strange demands. They are asking for one and a half cents of land for adivasis and patta for housing rights. The housing scheme does not apply to tribals who have been traditionally living in forests and what are the tribals to do with one and a half cent of land? 'A' also runs a tea estate here. In the beginning, they gave tea seeds to tribes and found that the project worked quite well. Now, they claim on being organic, and export it to Norway, Sweden. It is not even organic. . . . Then they sought funds saying that the tribes are cultivating tea and hence need an estate of their own. They got German funds and bought an estate. It's about 150 acres and is supposedly for the Paniyas. 'A' runs it and the tribes work there. Everyday, each woman worker is expected to pluck 20 kgs of leaves. Even if 1 kg is less, the worker will get only half the salary for the day. The other estates are better than this. The NGOs will not oppose any action of the government strongly as they cannot survive without their Foreign Contribution Regulation (FCR) funding for which they need to be in the good books of the government. . . . 32 lorries of timber go out of the forest everyday and they will not do anything. But they will go and report about people collecting firewood from the forest. We called up and informed them about a company which felled about 1,200 trees. No action was taken. You cannot sit comfortably in Chennaior Ooty and keep talking about how people are taking water, cultivating,

> collecting firewood and other produce. How is collecting firewood an issue at all? How will you save the forest? And the rest of us who do take up these issues are branded as terrorists. An interview in a regional newspaper mentioned that 12 panthers had died here. These animals had died within estate compounds. The impression given is that the tribals poison and kill the animals in the forest to make soup . . . that if the people continue to live within the forest, all the animals will disappear. They never take action on those who are actually responsible for such crimes. They don't even come down to the areas. These statements are made from kilometres away. They don't realise that it is because of the tribals and other people that the forests have been saved.

The excerpt highlights the domineering role of NGOs in forest areas, and also the role they play in selectively upholding the model of human-wildlife conflict to target tribal communities. It reveals the vested interests of NGOs whose stated intention is environmental conservation. The excerpt only reiterates the need to re-examine environmental agendas as modes of governance that have contributed to ensuing conflicts in tribal areas.

The following excerpt is of a tribal activist, a Kurumba, from Pandalur, Tamil Nadu. The main tribal communities in Pandalur are the Mullu Kurumba, Betta Kurumba, Jenu Kurumba, Paniya and Kattunaicka. The other communities include Nayars, Theeyars, Christians and Muslims from Kerala, Chettiars from Karnataka and dalits from Sri Lanka. Most of the Sri Lankan dalits and Paniyas are plantation labourers.

> The main reason that the adivasi people are not able to move forward is because of the NGOs. If one wants to serve, it is not an issue. But all actions of the NGOs are concerned with doling out money. There is one organisation here that has given people money for doing anything and everything, so much so that people will only listen either if the NGO instructs, or if you have given them money.
>
> In Pandalur, there is NGO 'C' which has been working in the area for a long time now. When their funders come from abroad they will give a long spiel about the kind of work they do. The adivasis do not know English and sometimes do not even understand Tamil. So what they tell you is what you know. But there is only 0.5 per cent Paniyas who have completed their 12th standard. In Kotagiri, many Paniyas have been taken by Malayali estate owners to work in ginger plantations. Here also they are paid the same wages of Rs 150 per day. But in the plantations, there is access to cheaper liquor. And it is exciting for them as it is a new place. The Paniyas are brought from Wayanad and also from Pandalur. There have been many cases of death in those plantations; six men from

> Pandalur have died in 5 years. They did not even bring back the body at times. They present suicide as the cause of death in most cases. But how can so many people commit suicide. One man died about 6 month ago. We are waiting for his post-mortem report.
>
> But there is a lot of underhand dealing between the estate owners and the police. We asked the NGOs to help out the community in this matter. None of them came forward.

The excerpt illustrates how the plantation economy dominates the region and the lives of local tribal groups. The NGOs mentioned by the Pandalur activist were all established after 1985. Each of these NGOs is involved in various activities: conservation, health, livelihood, education, social awareness through community radio and so on. One of the NGOs referred to in the excerpt owns a tea estate which was bought with the help of a British donor. It is also involved in other international projects. The patronising tone in the following quote from its website reeks of the politics highlighted by the Pandalur activist.

> An intangible gain (of starting the tea estate) is that local adivasis see that their prestige, their image in the area has improved because the adivasis now own a plantation, call the shots, are not anymore lowly unskilled labourers . . . self-esteem and confidence has grown. 'Few of our people stagger around drunk . . . Before it was just an adivasi drunk'. Now it's Radhakrishnan or Kumaran or Chimbaran not merely a drunken Paniya lying in the gutter. So we've come a long way.

The NGOs, clearly viewed as outsiders by tribal communities, now dominate governance in forest areas. They have a strong presence in various fields of development and are perceived as powerful interlocutors in the ongoing conflict over forest administration as well as conservation. However, they have only widened and further complicated the gap between the people and government departments. Supported by international and national funding, these NGOs form an integral part of the political economy of the Nilgiris. Their colonial approach towards tribes is evidenced in the way they have sought to quell resistance among local tribes, integrate tribes into various programmes and through the lure of money. The focus seems to be on assimilating tribes into the plantation economy that dominates the region, rather than seeking them out as partners in evolving a programme for conservation in the Reserve.

Conclusion

The 12 August 2015 report presented in the Rajya Sabha of the Parliamentary Standing Committee on Science and Technology, Environment and

Forests, evidences the analyses presented in this chapter. The committee members held a meeting on 17 July 2015 with NGOs/civil society members/experts to comprehend the environmental issues of Ooty at Udhagamandalam (Rajya Sabha, Parliament of India 2015). Most of the organisations that were part of the deliberations were precisely the ones identified by the local activists for their complicity with the state departments in sustaining the politics against local tribal communities. Based on its interaction with and suggestions of organisations present in the meeting, the committee recommended a sanction of 23.50 crores for relocating people from core and buffer areas of the Mudumalai Tiger Reserve and special financial assistance from the central government to the state government for vacating land for the elephant corridor. The committee only reiterated the model that has been the bane for the local tribal communities, i.e. man-animal conflict, and the need to reduce it by clearing human habitations within the reserves.

The article set out to illustrate government practices in the NBR and the new forms of bureaucratic encumbrances introduced in the region. These government practices configure the politics of conservation in the NBR. Various stakeholders have capitalised on the prejudice entrenched within environmental agendas and projects against tribes and on the question of human-wildlife cohabitation and coexistence. The question of environmental conservation is therefore highly political in nature, both in terms of who gets excluded and who stands to gain from it. Dominant stakeholders dictate conservation practices within NBR that have contributed to environmental degradation and undermined the biodiversity of the region. There is an urgent need to demystify the environmental model as a mode of governance and unravel its politics that has systematically marginalised the interests of local tribal communities.

Notes

1 This chapter was originally published as Ritambhara Hebbar, 'Nilgiri Biosphere Reserve: Reflections from the Field', *Sociological Bulletin*, Vol. 67, Issue 3, pp. 302–316. Copyright 2018 © Indian Sociological Society. All rights reserved. Reproduced with the permission of the copyright holders and the publishers, SAGE Publications India Pvt. Ltd, New Delhi. I would like to thank my research assistant, Ms. Lakshmi Premkumar, for her contribution to my research on the Nilgiris. I also want to thank Ms. Soummya Prakash, my PhD scholar, who assisted me in collating the references for this chapter.

2 Acronym for International Union for the Conservation of Nature and Natural Resources.

3 The proposal identified 14 scientific research institutions, 11 all India survey organisations, 16 universities, seven state government organisations and more than nine voluntary organisations in the region.

4 *Patta* is a title deed to a property.

5 Officially known as the Scheduled Tribes and Other Traditional Forest Dwellers (Recognition of Forest Rights) Act 2006, the act recognises right 'to hold and live in the forest land under the individual or common occupation for habitation or

for self-cultivation for livelihood by a member or members of a forest dwelling Scheduled Tribe or other traditional forest dwellers' (Ministry of Law and Justice 2006). While there have been concerns raised over its efficacy, the Act has provided some hope to forest dwellers in South India who contend with threats of evictions from the Forest Department.

References

Centre for Ecological Sciences. (1986). *Scientific programme for the Nilgiri Biosphere Reserve: Proposal for an action plan*. Bangalore: Indian Institute of Science.

Daniels, R. (1996). *The Nilgiri Biosphere Reserve: A review of conservation status with recommendations for a holistic approach to management India* (Working Paper No.16). UNESCO. South-South Co-operation Programme for Environmentally Sound Socio-Economic Development in the Humid Tropics. Retrieved from http://unesdoc.unesco.org/images/0011/001137/113753Eo.pdf

Department of Environment. (1984). *Demarcate the various zones of the Nilgiri Biosphere Reserve*. New Delhi: Government of India.

Gadgil, M., Nair, S. S. C., and Sukumar, R. (1986). *Scientific programme for the Nilgiri Biosphere Reserve: Proposal for an action plan* (CES Technical Report No. 37). Bangalore: CES, IISC.

Hockings, P. E. (1989). *Blue Mountains*. New Delhi: Oxford University Press.

Kothari, A., Suri, S., and Singh, N. (1995). Conservation in India: A new direction. *Economic and Political Weekly*, 30(43), 2755–2766.

Ministry of Law and Justice. (2006). *The scheduled tribes and other traditional forestdwellers* (Recognition of Forest Rights Act 2006). Retrieved from http://tribal.nic.in/WriteReadData/CMS/Documents/201306070147440275455NotificationMargewith1Link.pdf

Parthasarathy, J. (2007). *Problems of indebtedness: Tribal experience in Nilgiri district*. Udhagamandalam and Ooty: Tribal Research Centre and Hill Area Development Programme.

Rajya Sabha, Parliament of India. (2015). *Environmental issues of the Nilgiris*. New Delhi: Rajya Sabha Secretariat. http://164.100.47.5/newcommittee/reports/EnglishCommittees/Committee%20on%20S%20and%20T,%20Env.%20and%20Forests/268.pdf

Raman, K. R. (1986). Plantation labour: Revisit required. *Economic and Political Weekly*, 21(22), 960–962.

UNESCO. (1974). *Report of the task force on criteria and guidelines for the choice and establishment of biosphere reserves*. Washington, DC. Retrieved from http://unesdoc.unesco.org/images/0000/000098/009889EB.pdf

Venugopal, D. (2004). Development-conservation dilemma in the Nilgiri Mountains of South India. *Journal of Mountain Science*, 1(1), 74–80.

World Conservation Monitoring Unit. (1988). *Directory of the proposed Nilgiri Biosphere Reserve draft: Protected areas data unit*. Cambridge. Retrieved from https://ia800306.us.archive.org/6/items/directoryofpropo88atki/directoryofpropo88atki.pdf

6

NEW COASTAL CLAIMS AND SOCIO-LEGAL CONTESTATIONS IN MUMBAI

Artisanal fishers and the problematic of the urban environment

D. Parthasarathy and Hemantkumar A. Chouhan

Introduction: new coastal claims and contestation

This chapter draws from ongoing projects that seek to understand the various ways in which new geographies of law as well as their wilful violations and amendments with questionable motives result in contestations over the coastal commons in the Mumbai region. India's coasts are undergoing a momentous process of socio-economic change that involves many new and competing claims on space and resources. This has been accompanied by the development of new governance arrangements based on spatial zonation, which have now become the subject of intense societal debate. The artisanal fishing population of India has especially been worst affected, and they have responded both institutionally and organisationally to the set of new claims that are being imposed. Fishing is undeniably among the oldest economic activities prevailing along the coast of India's mainland, and constitutes a chief source of livelihood and basis for social organisation even in highly urbanised agglomerations such as the Mumbai Metropolitan Region (MMR). The number of active sea-going fishers in India is currently estimated at approximately 1 million, with the total number of people depending on fishing for a livelihood (fishers, traders, processors, and dependents) estimated at 25 million. Their reactions to processes and projects such as expanding coastal cities, Special/Coastal Economic Zones (SEZs), new ports/harbours, thermal/nuclear power plants, aquaculture farms, physical infrastructure, and coastal tourism have been varied. Our ongoing research attempts to investigate the institutional patterns that emerge along the shorelines of the various coastal states, as a response to the new coastal claims that initially encroach upon and ultimately displace coastal fishing communities from accessing the commons that are the basis for life and livelihood for tens of thousands of households in this region.

The new coastal claims exacerbate the diverse pressures on fisher livelihoods and survival as well as coastal commons that are already under severe stress due to coastal erosion, changing tidal patterns, loss of biodiversity, and ecological degradation, in a context of competition and conflict over fishing grounds involving fishers from other parts of Maharashtra and India, as well as global commercial fisheries. Thus fisher rights to space and resources are being threatened by a multitude of competing claims caused by industrialisation, urbanisation, climate change, etc. Amidst these vulnerabilities and insecurities, the role of environmental legislation such as the Coastal Zone Regulation Act and its various amendments have played multiple and often contradictory roles in redefining access to and the use of coastal commons. From a sociological perspective, the role of the state, inequalities related to class, caste, and livelihoods, economic growth imperatives, and environmental transformations impact the coastal commons and create new contestations even as older conflicts remain unresolved. Fisher use of marine and terrestrial space is thus being contested and reshaped in interaction with new claimants and their interests. The sizeable artisanal fishing population of the region bears the brunt of these impacts even as they receive little or no support from the state and the legal structure in contesting these new claims.

Historically, coastal fishers the world over depend on common pool resources (Garcia et al. 2014). They depend on marine species that are mobile and difficult to subject to property regimes. It is only recently, especially in the West, that private property in the form of individual transferable quotas has been introduced to fisheries, though informal and customary norms and practices do exist in the Indian context as well. Coastal fishers have historically made use of previously unclaimed and unused common pool resources that are unfenced. This also applies to the coastal terrestrial fringes that fishers occupy, for instance mangroves, mudflats, and wetlands were previously not of much relevance to other social and economic actors. Hence, such marine and terrestrial zones have been grouped in the legal/normative category of 'the commons', of relevance mainly to commoners, such as fishers (Ostrom et al. 2002).

Over time, fishing communities in the Mumbai region as in other parts of the world have developed complex forms of customary law to regulate the use of the commons, both marine and terrestrial (Bavinck et al. 2013). The state has generally tolerated this usage within the framework of a legal order that on the one hand affirmed its jurisdiction over marine and terrestrial territory, but simultaneously also affirmed their marginal status. In Maharashtra, the legal status of terrestrial commons along the coast were governed by the identification of coastal villages as *gaothans* or *koliwadas*; ownership and management rested with the village or community rather than higher government agencies, and in fishing villages areas were demarcated for fisheries related activities. There were various restrictions on urban development in these *gaothans*. Marine waters and fishing grounds were demarcated usually through negotiations between neighbouring villages, districts, and regions, with customary laws governing fisheries within each boundary, some of which have, in recent years, gained formal legal status.

This chapter is to be seen as a contribution to the reframing of the concept of environment which also subsumes a reframing of territorial units such as the rural, the urban, and the regional. Sustainable coastal environment management requires an appreciation of the role of natural resources and ecosystems for livelihoods, disaster mitigation, and human well-being, but also a recasting of the ways in which we theorise and plan for the urban, including economic development in urban areas. Further, as incidents of flooding in rural and urban areas increase in number and intensity, the need for regional level hydrological planning becomes evident, a process which requires an abandoning of conventional territorial-administrative boundaries, in favour of a more ecological understanding of regional planning from a development, disaster mitigation, and sustainability perspective.

In line with changes the world over (Smith 2000), the Mumbai coast is now subject to coastal squeeze (Pontee 2013) as well as development pressure. Coastal cities such as Mumbai with high population densities and high rates of economic growth are expanding, and spaces in the coastal commons in the entire region are now sought for purposes of tourism development, recreation, aquaculture, industry, mineral exploitation, ports, power plants, roads and expressways, and environmental conservation (Lakshmi et al. 2012). Sea level rise, coastal erosion, and coastal flooding have also put human security issues on the agenda (Parthasarathy 2009). In response to increasing demands from environmentalists and resource dependent communities, the government of India has been formulating and implementing new and detailed coastal management plans. The CRZ notifications of 1991 and 2011 (with a large number of amendments in between) have been heavily contested by the fishing population and environmental groups (Mathew 2008; Sharma 2011), even though they were designed precisely to protect coastal communities, environments, and habitats. As in other parts of the world (Sevilla-Buitrago 2015; Sugden and Punch 2014), the coastal commons in the Mumbai region are now the object of severe contention.

In seeking to better understand the current contest for coastal space in the Mumbai region, and its effects on the fishing population, our ongoing research (and this chapter) centres on the erstwhile commons and considers socio-legal changes occurring since the liberalisation of the Indian economy in 1991. It studies the legal, institutional, and governance aspects of coastal contestations. A key objective was to understand the role that law and more generally normative systems play in effecting, countering, and negotiating new claims to coastal commons in the region. Following

Benda-Beckmann and Benda-Beckmann (2006), we suggest that claims to space and resources are fought out in the legal realm, and that an analysis of normative issues throws light on the dynamics of societal change responding to coastal claims and attendant socio-economic transformations. A comprehensive assessment of the role of CRZ laws that facilitate arrangement and re-arrangement of claims to coastal commons is a significant objective of this study, something that has not yet been attempted rigorously despite over two and a half decades of existence of CRZ rules. A substantial and long-term objective of the study is to draw attention to the issue of how CRZ balances out customary claims with new development imperatives, and the increasing demand for coastal common spaces with environmental sustainability. Another important aim was to analyse how various forms of law were used to promote particular claims over coastal space, and to assess fisher responses reflected in everyday forms of resistance, collective protests, and citizenship claims made through courts, tribunals, and other legal avenues. It is observed that inequalities (gender, ethnicity, class, power) intermediate fisher responses to the adverse effects of new coastal claims on livelihoods and coastal environments. As such an alternative understanding of social inequalities and legal transformations with reference to coastal commons becomes an essential aspect of reframing the environment and struggles around the environment, especially as both the state and market forces redefine the environment arbitrarily to enhance control and advance the process of accumulation by dispossession.

Through documenting and describing a series of encroachments on coastal commons, identifying the impacts of new claims on coastal commons, and tracing the implications of CRZ norms and violations on coastal communities, this chapter explains the uses and misuses of law that constitute important strategies for making, contesting, and negotiating old and new coastal claims.

Coastal claims, encroachment on commons, and marginalisation of fishers in the Mumbai region

There are more than 24 Koliwadas and more than 88 Gaothans in the Greater Mumbai region. Among them more than 16 Koliwadas and 23 Gaothans are in the CRZ zone. Most are located on the seashore or in nearby areas. Although each Koliwada is unique, they share common characteristics and problems. Pollution, lack of basic amenities, declining catch, increasing costs, inadequate housing and declining space for trade related activities are the issues affecting most Koli families (Warhaft 2001). On the other side there is landward encroachment by migrants and CRZ violations. The main problems in the MMR coastal region emerge from changing land use patterns, residential and industrial water supply and waste disposal, transportation-related air, soil, and noise pollution, coastal marine pollution, and depletion of important coastal habitats like wetlands, mudflats, salt pans, and mangroves. Parts of the coastal zone of MMR have also become increasingly susceptible to human-induced environmental stresses and economic damage by natural geophysical factors such as erosion, siltation, and coastal flooding. The waste generation and disposal pressures due to domestic and industrial activities have also contributed to the deterioration of coastal marine water quality and coastal fisheries. The MMR coastal region is the key commercial, financial, and industrial region of India with around 9,000 small, medium, and large (public and private) industries ranging from chemicals, fertilizers, iron and steel, oil refineries, and thermal power operating here. Industrial pollution in the MMR has not been fully assessed, but the main sources include gaseous emissions, solid and liquid wastes, toxic and hazardous wastes, and construction debris (Murthy et al. 2001).

The Kolis of Mumbai have been confronted with a seemingly unstoppable barrage of neoliberal development – projects, programmes, and policies stemming from transnational, national, regional, and local conditions. The Kolis, the vast majority of whom practise small-scale artisanal fishing techniques, are implicated in a globalised seafood export economy. Nationally, central government agencies continue to promote destructive policies rooted in the ideology of modernisation and orchestrated by plans that have resulted in the benefits of increased production flowing to people who have never set foot on a fishing boat. At a time when fishers still lose their lives at sea for want of a simple lighthouse, state governments authorise and support gigantic port projects for international trawlers, whose mass fishing techniques plunder the oceans leaving one environmental catastrophe after another in their wake. And at the local level, development projects and urbanisation threaten seaside settlements in almost every Koliwada (Koli hamlet) in the city (Warhaft 2001). Marine pollution from industrial and domestic loads as well as hydrocarbons and tar deposits led to degradation of most of the beaches and beach waters around Mumbai. Mangroves are the most important component of the coastal ecosystem, keeping the shoreline intact against tidal currents by preventing soil erosion. They also provide a habitat for several marine species, including birds, shrimp, and fish (Murthy et al. 2001). Parthasarathy notes that during the 2005 floods, mangroves and salt pans in the eastern suburbs mitigated the scale of disaster in some areas. Destruction of mangroves was a key element of flooding in the western suburbs. Dense mangroves are important for ecological and environmental services, but they are destroyed to make way for 'development' (Parthasarathy 2011). Unplanned development and high population density impose severe restrictions on resources and lead to conflicts among coastal stakeholders and new claimants (Warhaft 2001). The resultant process of marginalisation and the loss of livelihood among the artisanal fishing communities have transferred them to the category of 'ecological refugees' from being environmentally sustainable protectors of coastal habitats. However, despite such transformations, there is an almost complete absence of effectual policies measures to address these social and environmental crises.

It is against these developments that a series of case studies were/are being undertaken in the Mumbai region to document and describe the declining access to and resource degradation in the coastal commons. From the early 1990s to the present, commons used by fishing communities for parking their boats, landing grounds for boats, spaces for drying and marketing fish, and repairing of nets have been gradually encroached upon through legal and illegal means, often in violation of or through surreptitious amendment to the CRZ laws. Three types of new claims impacting on terrestrial and marine commons are identified: (1) private appropriation of commons and commoned resources by industry, public sector, and real estate, infrastructure, and service sector; (2) socialisation of commons reflected in conversion of coastal spaces into zones that provide municipal, economic, public, and ecosystem services; and (3) destruction of the environment and ecological degradation in the commons resulting from dumping large amounts of pollutants, garbage, and debris by the state and by private players.

From the Esselworld and Water Kingdom recreation spaces in Goregaon, to the new water sports centre in Girgaum, fishers are continuously being restricted in their activities through exclusion from their erstwhile coastal spaces. They are also being excluded from other fisher spaces owing to security issues such as coastal commons near oil drilling sites, refineries, atomic research centres, and power plants. In many of these cases, beyond the actual encroachment, nearby spaces commoned by fishers have also become degraded due to related activities. For instance, mangroves in Gorai, Sewri, and Mahul – which are an important breeding ground for fish and a source of biodiversity – have become polluted due to dumping of chemical effluents, storage of coal, and spraying of chemicals. Many of these are classified as CRZ-I, and are legally off bounds for non-fisheries-related activities.

The Bombay High oil wells in the deep sea and drilling for oil has encroached upon and disturbed the fishing grounds, apart from restricting fishing in these areas. The proposed new airport and SEZs in Navi Mumbai have already severely trespassed upon coastal commons, and created new disturbances to the hydrology of coastal wetlands and ecosystems. Estuary zones that are part of fishing-related activities and encroachment on which affect tides and currents, thus affecting fish movement, have been intruded into and reclaimed by infrastructure projects such as the Bandra Worli Sea Link project. Further the Sea Link project displaced two koliwadas, affecting the livelihoods of thousands. Mangroves have been hacked to make way for exclusive apartments for the rich and for large commercial projects. New government plans to build more bridges, coastal roads, trans-harbour links, transport projects, memorials, and tourism projects further enhance the threat of coastal degradation and marginalisation of the fishing community.

Most koliwadas in Mumbai, which have traditionally enjoyed legal protection, are now invaded and reclaimed for various developmental projects. The Cuffe Parade fishing community successfully protested a floating five-star hotel that endangered their livelihood and threatened the displacement of more than 1,000 families in their koliwadas in the central business district of Nariman Point. However, this limited success follows a long and sustained process of gradual takeover of fishing grounds in the same area by government and private housing projects. Following the trend of bourgeois environmentalism (Baviskar 2011), the Cuffe Parade Residential Association constructed a garden by destroying 60 square metres of mangrove forest in the area.

Bandra, Mahim, and Versova creeks have been polluted through regular dumping of construction debris and flow of untreated effluents into the creek and sea. This has severely affected fish catch, as diverse species of fish are seen dying in large numbers. Beyond Mumbai city, in the larger metropolitan region, many new projects have laid claim to coastal commons and more are on the anvil. These include a large number of tourist, recreational, and hospitality projects in Palghar and along the Vasai-Virar belt, a range of commercial commercial complexes in Andheri, a large port in Wadhvan, and a thermal power station in Kelve. In the Thane-Mulund belt 134 acres of mangroves have been destroyed for an SEZ.

As in several other cities, the crisis of solid waste management is resolved through encroachment and destruction of the commons. The battle over the Kanjur Marg municipal dumping ground in violation of CRZ norms best exemplifies this. Expressing the complex historical relationship between categorisation and management of waste, commoning, and socialisation/privatisation of the commons (Gidwani 2013), the state successfully re-classified and converted 'wet land into 'waste land' in a CRZ area, destroying flora and fauna in the eastern coastal belt and adversely affecting both livelihoods and health of adjacent communities (Sharma and Parthasarathy 2015). In many locations, CRZ notifications have been relaxed, violated, or amended to pave the way for commercial exploitation of coastal zones, including in CRZ I – the most ecologically sensitive zone.

Fisher resistance to new claims: challenges and possibilities

From an environmental justice perspective, the continuous marginalisation of Mumbai's fishers constitutes an example of geographic inequity. Already buffeted by depletion of resources due to commercial over-fishing and climate-change-related livelihood impacts, artisanal fishers in the Mumbai region are facing further marginalisation due to new coastal claims and encroachments in CRZ notified areas. As neoliberal urbanisation and speculative capital make new claims and demands on urban coastal zones with implications for coastal degradation and livelihoods, fishing communities are organising themselves and fighting back to defend their community rights and restore marine ecologies supported by civil society organisations and environmental activists. A key instance is that of the Uran fishers, who succeeded in obtaining a favorable judgment from the National Green Tribunal not just in terms of a fair compensation, but also – for the first time – recognition of and compensation for commons that are destroyed and they are excluded from. This constitutes a significant victory for fishworker associations and environmental NGOs. However, such examples are rare, even though its significance in the long term for fisher struggles should not be underestimated. The judgment itself has been challenged by the state, and the fishers are still fighting for their rights in this case. This is not withstanding the neoliberal thrust of environmental legislation, their amendments, and violations – the most evident being those pertaining to coastal regulation zones.

Fishworker organisations such as National Fisher Forum (NFF) and other agencies have been continually critiquing the CRZ policy as well as its several amendments and large-scale violations. Despite their limitations, and overall lack of success, such protests and actions by fishers achieve small successes in terms of sustaining media and environmentalist attention, drawing limited and short-term political support, and ensuring minimal success in preserving fisher spaces in coastal zones for the short term. Continual protests and everyday resistance have become the norm, affecting livelihoods and coastal environments, but ensuring that there is no easy defeat by state and market forces out to usurp coastal lands for new developmental projects. Both nationally and in the Mumbai region, fishers and environmentalists, despite not always being in agreement about strategies and goals, have maintained pressure against state failure to implement CRZ norms in letter and spirit, and against CRZ violations.

James Scott (1985) defines subordinate resistance as,

> any act(s) by member(s) of a subordinate class that is or are intended either to mitigate or deny claims . . . made on that class by superordinate classes (for example, landlords, large farmers, the state) or to advance its own claims . . . vis-à-vis those superordinate classes.
> (Scott 1985)

This definition broadly covers resistance that is individual and collective, material and symbolic, failed and successful.

The aim of everyday resistance is to test the limits to practices and customs, rather than drastic change, where both subordinate and dominant groups are constantly trying to hold the advantage of everyday relations (Scott 1985). Likewise, although communities such as the fishers face resistance problems similar to those of peasants studied by Scott, conservation regulations such as CRZ norms are framed and governed by the desire or greed to create more capital accumulation. Restricting coastal spaces for artisanal fishing and allowing big trawlers to operate in the same spaces creates new spaces of resistance but also constrains fishers from different directions. Power relations operate here as different players deal with restrictions and environmental degradations arising from new coastal claims, climate change, environmental pollution, and depletion in diverse ways. These exacerbate livelihood insecurity for the fishing community. As George Holmes (2007) points out, politics plays a significant role in defining and delimiting resistance even as the state enacts new norms and laws to defend protected zones and their natural resources and removes dependent communities from these resources.

The process of resistance comes to be marked by a continuous set of divergent and many-sided conflicts, which fishers need to negotiate. In the case of CRZ-related resistance to state violations and implementation by amendment, a series of conflicts, developmental claims, and encroachments define the modes of resistance by fishers. Coastal communities and groups such as the fishers and other coastal populations enjoyed customary or traditional rights to exploit resources and to fish in nearby coastal areas. The central and state governments have severely impacted these customary practices taking away communitarian controls (Rodriguez 2010). Fishing communities in India and the Mumbai region have struggled for greater control over the seas and resource management, struggles which have been directed both inward as well as against the state. The history of change in maritime societies has witnessed conflict between small fisher communities and the large mechanised sector, especially with respect to competition over resources or, in other words, to the 'common' space of the sea (Ram 1991). Fisher communities, particularly the artisanal communities, find themselves most impacted by this alienation, as the development drive has left them marginalised (Vivekanandan 2007). The patterns of resistance that emerge are transformed by the type of conservation measures being enforced, the methods of enforcement, the new developmental claims, and the social contestation around these that involve a range of poor, middle class, and elite groups.

Coastal fishing villages in the Mumbai region are facing a double jeopardy from conservation interests and development imperatives – with fishers being marginalised and alienated from their workplaces and residential areas because of conservation needs and development projects. The Swaminathan committee recommended that 'future policies for coastal area management must reverse these trends and find approaches to conserve and protect vulnerable ecosystems and secure livelihood of all these habitats'. Owing to large-scale state and policy failure in giving force to this recommendation, fishing communities in the region have resorted to several formal and informal strategies – legal and political – in battling further marginalisation and reclaiming coastal zones which have both livelihood and long-term environmental consequences. These include protest marches, forum shopping by appealing to diverse judicial bodies including the courts and green tribunals as well as human rights bodies, and the use of constitutional measures relating to decentralisation. All of these express a reframing of the environment not just from a natural resource, livelihood, economic growth, or urban planning perspective, but point to a process whereby emerging legal and normative regimes situate environmental law and governance as a 'semi-autonomous social field' (Moore 1973), and need to be studied from that perspective.

References

Bavinck, M., D. Johnson, O. Amarasinghe, J. Rubinoff, S. Southwold and K.T. Thomson (2013). From indifference to mutual support: A comparative analysis of legal pluralism in the governing of South Asian fisheries. *European Journal of Development Research 25* (4): 621–640.

Baviskar, Amita. (2011). Cows, cars and cycle-rickshaws: Bourgeois environmentalism and the battle for delhi's streets. In Amita Baviskar and Raka Ray, eds. *Elite and everyman: The cultural politics of the Indian middle classes* (pp. 391–418). New Delhi: Routledge.

Benda-Beckmann, F. von and K. von Benda-Beckmann (2006). The dynamics of change and continuity in plural legal orders. *Journal of Legal Pluralism 53–54*: 1–43.

Garcia, S., J. Rice and T. Charles (Eds.) (2014). *Governance of marine fisheries and biodiversity conservation: Interactions and co-evolution.* Oxford: Wiley-Blackwell.

Gidwani, V. (2013). Six theses on waste, value, and commons. *Social & Cultural Geography, 14*: 773–783.

Holmes, G. (2007). Protection, politics and protest: Understanding resistance to conservation. *Conservation and Society 5*: 184–201.

Lakshmi, A., A. Schiavina, P. Banerjee, A. Reddy, S. Mandeen, S. Rodriguez et al. (2012). *The challenged coast of India.* PondyCAN in collaboration with BNHS and TISS.

Mathew, S. (2008). Coastal management zone: Implications for fishing communities. *Economic & Political Weekly 43* (25), June 21: 17–24.

Moore, Sally Falk. (1973). Law and social change: The semi-autonomous social field as an appropriate subject of study. *Law and Society Review*: 719–746.

Murthy, R.C., Y.R. Rao and A.B. Inamdar. (2001). Integrated coastal management of Mumbai metropolitan region. *Ocean & Coastal Management 44* (5–6): 355–369.

Ostrom, E., T. Dietz, N. Dolšak, P.C. Stern, S. Stonich and E.U. Weber (2002). *The drama of the commons.* Washington, DC: National Academy Press.

Parthasarathy, D. (2009). Social and environmental insecurities in Mumbai: Towards a sociological perspective on vulnerability. *South African Review of Sociology 40* (1): 109–126.

Parthasarathy, D. (2011). Hunters, gatherers and foragers in a metropolis: Commonising the private and public in Mumbai. *Economic and Political Weekly 46* (50): 54–63.

Pontee, N. (2013). Defining coastal squeeze: A discussion. *Ocean & Coastal Management 84*: 204–207.

Ram, Kalpana. (1991). Mukkuvar women: Gender hegemony, capitalist transformation in a South Indian fishing community. New Delhi: Kali for Women.

Rodriguez, S. (2010). *Claims for survival: Coastal land rights of fishing communities*. Dakshin Foundation: Bangalore.

Scott, J.C. (1985). *Weapons of the weak: Everyday forms of peasant resistance*, New ed. New Haven, CT and London: Yale University Press.

Sevilla-Buitrago, A. (2015). Capitalist formations of enclosure: Space and the extinction of the commons. *Antipode 47* (4): 999–1020.

Sharma, C. (2011). CRZ notification 2011: Not the end of the road. *Economic & Political Weekly 46* (7): 31–35.

Sharma, S. and Parthasarathy, D. (2018). Urban Ecologies in Transition: Contestations around Waste in Mumbai. In Jenia Mukherjee, ed. *Sustainable Urbanization in India* (pp. 207–223). Singapore:Springer.

Smith, H.D. (2000). The industrialisation of the world ocean. *Ocean & Coastal Management 43*: 11–28.

Sugden, F. and S. Punch (2014). Capitalist expansion and the decline of common property ecosystems in China, Vietnam and India. *Development and Change 45* (4): 656–684.

Vivekanandan, E. (2007). Impact of climate change on marine fisheries. *Seafood Export Journal 37* (4): 5–9.

Warhaft, S. (2001). No parking at the bunder: Fisher people and survival in capitalist Mumbai. *South Asia. Journal of South Asian Studies 24* (2): 213–223.

7

BONDS THAT DIVIDE

Urbanisation and the erosion of the commons

Hita Unnikrishnan, B. Manjunatha and Harini Nagendra

The lives and livelihoods of many people within rapidly urbanising India are heavily dependent upon its vast reserve of common pool resources (Narain and Vij 2015). Common pool resources include a group of resources, mostly community managed and in which the two important characteristics of subtractability and non-excludability are defining (Ostrom 1990). Subtractability refers to the fact that the withdrawal of resource units from such resources reduces the total amount available to other users, while non-excludability indicates that it is impossible to exclude users from such resources (Berkes et al. 1989; Ostrom 1990, 2007). Examples of common pool resources in India include lakes, ponds, step wells, grazing commons, sacred forests, village forests, cemeteries, and temple tanks among many others (Narain and Vij 2015). Many of these are steeped in historical significance and have been a part of the lifestyles of traditional communities for centuries (Agarwal and Narain 1997; Unnikrishnan and Nagendra 2016).

Along with the crucial role that they play in sustaining lives and livelihoods of dependent communities, common pool resources are also ecologically significant. They form important biodiversity hotspots and provide critical ecosystem services such as groundwater recharge, microclimate regulation and nutrient cycling (Kiran and Ramachandra 1999). As these resources are both socially and ecologically significant they form true social ecological systems, and because of this characteristic, it is quite impossible to separate either the social or the ecological component while making important decisions surrounding them (Unnikrishnan and Nagendra 2016).

Their criticalities to traditional communities notwithstanding, urban commons are severely threatened (Agarwal and Narain 1997). Increased state control over these resources since colonial times, conversion to private ownership regimes and physical transformations of community-based resources into built-up structures have largely distanced traditional communities from them (Agarwal and Narain 1997). In such a landscape where the social component of the social ecological system is largely alienated, these resources are further rendered vulnerable to threats such as pollution, encroachments and pollution.

Urban commons in the country are also threatened due to neoliberal regimes operating across the country. Neoliberalism has become an integral part of today's world, an essential prerequisite to understanding our interpretation of its functioning (Harvey 2007). It has become an important process influencing change in regimes of environmental governance, and has been implicated in increasing environmental risks (Prudham 2004; Castree 2010). Neoliberalism refers to the theory that proposes human well-being to be benefited by an institutional regime comprising of private property rights, individual liberties, 'unencumbered' markets and trade (Harvey 2007), and where the state makes available both the institutional framework as well as the market for such regimes. Yet, the fundamental question that is posed by critiques of neoliberalism across the globe is whose interests are represented within the neoliberal regime (Harvey 2007; Castree 2010). This chapter engages with this pertinent critique of neoliberalism, highlighting a scenario with respect to use and management of urban commons. We focus our attention on the lake based urban commons of the south Indian city of Bengaluru, and in so doing attempt to shed light on the question of who really owns these resources.

The south Indian city of Bengaluru, capital of the state of Karnataka, is globally renowned as the 'garden city' of India (Sudhira et al. 2007). It has garnered further attention in recent years for its contribution to the information technology sector of the country. Apart from these achievements, Bengaluru is also home to some very fascinating urban commons – the networked lakes that characterise the landscape, traditionally protected village forests (*Gundathopes*), temple tanks (*kalyanis*) and village grazing commons (*gomalas*) to name a few (Mundoli et al. 2014). Historically significant and integral to traditional village communities, they still have undergone drastic transformations owing to the rapid pace of urbanisation witnessed by the city. This chapter takes the case of lake commons in Bengaluru to illustrate the impacts of transformations in the nature of use and access on communities dependent upon them. We further link these changing community relations with changes in the ecological characteristics of the resources and underscore the importance of inclusive planning in a system heavily relying on citizen ecosystem stewardship for the management of natural resources.

Study area

The south Indian city of Bengaluru has always been a major urban settlement and a defensive stronghold since historical times. Evidence of prehistoric settlements and written stone inscriptions dating as far back as the 8th century attest to the presence of human habitation and civilisation in this city since ancient times (Annaswamy 2003). This long history of human presence and flourishment in any location devoid of major water resources such as land-locked Bengaluru is rather unusual. This oddity may only be explained when one takes into consideration the impressive, engineered, networked system of tanks or lakes that characterises the landscape (Rice 1905; Nair 2005). Built by making use of natural depressions in the topography, these lakes, some of which date back to even the 9th century (according to stone inscriptions dating to that period), once formed the water lifeline of the city (Rice 1905). In fact, their importance was so great that it was considered an act of great spiritual significance to have a lake built. Their maintenance and upkeep rested with communities (hence making them functional commons), and this too was accorded great significance (Buchanan 1807; Dikshit et al. 1993). Inscriptions record the most generous of blessings to those who followed rules related to the maintenance and upkeep of these water bodies, while predicting the goriest of misfortunes upon violators (Rice 1905). These inscriptions also record that communities took active part in the maintenance and management of lakes – this is evidenced by numerous records of land grants made to villagers in lieu of services made towards water bodies (Rice 1905). Lakes were important both for domestic activities such as bathing and washing clothes and vessels, but also for agricultural and industrial purposes (Buchanan 1807). Due to the fact that they were well maintained and protected by local communities, with bureaucratic support, these lakes eventually supported a wide variety of biodiversity, some of which were even economically important. They thus formed true social ecological systems, dependent on both their social and ecological components for proper functioning.

Over time, perhaps beginning with the rise of colonial rule in the city and the introduction of state control over former common pool resources (such as these lakes) that continues into the contemporary day, people gradually became distanced from these water bodies (Unnikrishnan and Nagendra 2016). In addition, the introduction of piped water supply to the city from the river Cauvery further pronounced the drastic decline of dependence on lakes, rendering them vulnerable to various threats such as pollution, encroachments and conversion into built-up structures. Today lakes within the city are mere shadows of their former selves, reduced drastically in numbers, size and presence within the city. Yet by no means are they any less important to the lives and livelihoods of communities who still exist in pockets of the rural even within the heart of the city. These people continue to derive sustenance and economic benefits from their dependency on the resource. Despite the importance of these water bodies in meeting the requirements of marginalised and vulnerable populations, these uses rarely fall within the purview of neoliberal regimes of governance currently operating within these spaces. As a result, several of these communities and their dependencies are excluded from lake development programmes – both state-led as well as collective – leading to their eventual alienation from the resource, rendering them even more vulnerable and marginalised. These effects of neoliberal regimes of resource governance are often rendered invisible in grands schemes of urban development and as such are important considerations that need to be highlighted, especially in the context of knowledge building for upcoming secondary cities that are characteristic of the developing global south. We therefore use the case of 20 lakes spread across the official boundaries of the city of Bengaluru, across both a gradient of size and urban cover, to illustrate the dependencies people build on these resources and also the ways in which they are alienated within processes of mainstream neoliberal urban development.

Methods

We used stratified sampling to select the study sites from a map of Bengaluru's lakes overlaid on the city's administrative boundary using ArcGIS. Care was taken to ensure that only extant lakes were depicted on this base map. Lakes were sorted out into sizes (small, medium and large) and random sampling was carried out within each of these subsets. Accordingly, 20 lakes were selected – seven small lakes, seven medium-sized lakes and six large lakes (which formed the entire subset of large lakes).

Data was obtained through a combination of field studies and interviews with elderly members of local communities. Multiple transect walks were conducted over two years around each lake in the study area to record the various ecosystem services derived from them. This data consisted of both direct visual observations of uses as well as evidentiary documentation of ecosystem services derived. In addition, detailed interviews were conducted with both ecosystem service users encountered at each lake, as well as elderly residents of the traditional village communities present around each lake. We used snowball sampling techniques to identify elderly members of communities. We asked questions aimed at getting information on former dependencies on lakes by communities and the changes they have seen over the past few years. Discussion also included the impacts of these changes on their lives and livelihoods. Over 15 in-depth interviews were thus conducted with ecosystem users and about ten in-depth interviews were conducted with elderly local villagers around each lake.

Results

Our results indicate that lakes are still dynamic spaces used by communities for a wide variety of purposes. While recreational activities of various forms (walking, jogging, exercising, playing music, angling and relaxing) remain the most visible of all uses of these lakes, a number of other provisioning and cultural ecosystem services are to be found practised around these water bodies. This diversity notwithstanding, many of these lakes have measures put in place that deliberately or unwittingly exclude these users from access to and appropriation of the resources. This section is therefore divided into three sections: (1) provisioning ecosystem services derived from lakes, (2) cultural ecosystem services derived from lakes and (3) modes of exclusion of traditional communities from access to ecosystem services from lakes.

Provisioning ecosystem services from lakes

Lakes within the city of Bengaluru provide a wide variety of provisioning ecosystem services to communities whose lives and livelihoods are dependent upon these services. Water from the lakes is used for the cultivation of crops – both agricultural and horticultural – especially on the banks of the lake. It is not uncommon to see lush paddy, millet or rose farms against an urban backdrop of flyovers and glass-covered high-rises within the heart of the city. Most farmers belong to communities for whom agriculture has been a traditional occupation that has passed down through several generations. The pumping of water from lakes for agriculture is also a common sight irrespective of the ecological condition of these water bodies. Water is often pumped into farms over a few kilometres away from the banks of the lake. In addition to the direct provisioning of surface water from the lake, groundwater from wells in close proximity to the lakes is often used for irrigation purposes. Agricultural produce from these fields are sold in local markets in the vicinity, sometimes even finding their way into the city's prominent marketplaces.

Apart from water for irrigation, lakes support livelihoods for fishermen living on their banks. In pre-independence eras, this occupation was hereditarily passed down to a group of people called the *Bestharu* (traditional fishermen communities). However, in the contemporary age, fishing is a tender regulated process, organised by the state fisheries department. Fishermen often display initiative in the care and maintenance of the fry let into the lake – sometimes even procuring fishmeal from other states as part of this process. The diversity of fish caught largely depends upon the fry let into lakes. As a fisherman in one of the studied lakes told us, "People want big fish from these lakes, but the lake in itself is not nutritious enough for the fish. So twice a year, we get two big truckloads of fishmeal from Tamil Nadu [a neighbouring state] and introduce it into the waters. It is hard work, but the fish grow to big sizes" (translated by authors from the regional language Kannada, which was used to conduct these interviews). In most cases, the fry of catfishes, labeo and tilapia are most commonly obtained from the fisheries department, and therefore harvested from the water bodies. Water from the lakes is also provisioned for commercial laundering of clothes, practised both by traditional laundering communities (*dhobies*) as well as migrants into the city. The washed clothes are dried on the banks of these lakes. Open wells near lakes are also sometimes used by these communities to practise their trade.

Women from communities near lakes are often seen harvesting the reeds growing in the shallow waters of the lakes. These are either used to feed their livestock or are sold to other livestock owners for money. The women also harvest green leafy vegetables growing along the banks of lakes. Fodder grass growing on the banks is an important nutritional source for livestock owners in nearby localities. These either are accessed by livestock owners practising pastoralism here or are collected by women for later use as fodder. These grasses are also harvested on larger scales along with certain types of floating vegetation for sale to livestock owners in distant places, as well as to a prominent zoo located in the city. Some lakes of the city are also important to local educational institutions in that they provide much needed laboratory specimens also used in various biological experiments. Organisms such as leeches and Chironomus larvae (important to zoology laboratories within educational institutions) are collected by suppliers along the banks of water bodies and sold for experimental purposes to educational institutions. On one of the lakes within the study area, a supplier told us, "These lakes are a great source to obtain these larvae – many laboratory suppliers and laboratory assistants from colleges all around the city come here to obtain them for their colleges."

In addition to these livelihood-enhancing or -supporting provisioning services, lakes are also important for various domestic activities, especially to the lives of low-income migrants into the city. Their dwellings are often close to lakes, and water from the water bodies plays an important role in enabling various domestic functions such as bathing and washing clothes and vessels. Village commons such as sacred forests situated next to lakes still provide much-needed shelter and shade to both nomadic communities as well as pastoralists taking their cattle out to graze. Certain root, tubers and green leafy vegetables growing on the banks of the lakes such as those of *Alternanthera sessilis* have perceived nutritive value especially in conditions of drought and famine and these provide a measure of food security in such conditions. In one such interview, we were told that,

> Onagane soppu (*Alternanthera sessilis*) is a plant that is found on the banks of the lakes. It helps to maintain hydration within our bodies and therefore, when water is less, or when food is scarce, we collect this vegetable and prepare it for consumption.

Several interviewees also recalled that this vegetable was an important source of nutrition, particularly during periods of drought and famine even during the time their ancestors lived and worked the land.

Cultural ecosystem services from lakes

Given their importance to sustaining lives and livelihoods, it is not surprising that lakes are sites of important cultural beliefs to lake-dependent communities. Many of these beliefs stem from people's dependencies on the water bodies and are therefore unique to each community. In general, each lake has a lake deity – often a feminine deity who is believed to protect associated villages from floods, disease and famine. This deity is believed to grant peace and prosperity, and protect both villagers and livestock from disease, drought and ruin. Offerings are made to these deities for various events and purposes such as marriage, general health, protection or cure from disease, in return for favours bestowed and in case of death. Each ritual is specific in the kind of offerings made and it is possible to deduce the ritual performed through the kind of offerings left behind. These deities are also worshipped in case the worshipper wants to inflict harm to anyone.

Apart from these individual belief systems, entire communities get together to worship lake deities, both on the basis of ecological events such as the overflowing of a lake due to an abundance of water as well as on the basis of their specific lake-based occupations. For instance, while the farmers perform the ritual for a good crop, washer folk do so to protect their unattended children as well as the clothes entrusted in their care (Box 7.1). Almost all of these rituals involve a procession around the villages connected by the lake, festivities, ritualistic sacrifices, feasting and the floating of numerous lit oil lamps on the surface of the water body.

Box 7.1 Some of the cultural traditions around lakes as practised by different occupational groups

Theppotsava: a ritualistic procession conducted by both farmers and shepherds of the community. This is normally celebrated when lakes overflow during the monsoons, indicating a bountiful harvest and prosperity. A *theppa* or float is constructed, and a statue of the presiding deity (often feminine) placed upon it. After the performance of ritualistic obligations (and an animal sacrifice), the float is carried to the centre of the water body and dispersed. Numerous lit oil lamps made of rice flour are also set afloat on the water body as part of this festival.

Gange Pooje: An annual worship of lakes that involved setting afloat lit oil lamps onto the water body to celebrate the sacredness attributed to water bodies.

Uppudyaavara pooje: Annually celebrated by commercial launderers (*dhobies*), festivities involve the creation of numerous mud representations of the patron deity Uppudyaavaru on the banks of the lake and a ritualistic procession around the community, followed by offerings made to the deity and communal feasting. Blessings are invoked for (a) protection for children left unattended by parents on the banks of the lake while washing clothes and (b) protection for the clothes entrusted to the communities by their employers.

In recent years, ecological changes have meant that certain user groups cannot use the water body anymore. For instance, in one of the lakes we studied, we were told that the quality of water has so deteriorated in the recent years that farmers are unable to use it for their crops. As a result, agriculture is no longer practised in the locality. However, pastoralists still use the banks of the lake to graze their cattle and sheep. It is interesting that in this instance, while farmers have alienated themselves from the lake in question, even abandoning rituals they used to practise, their cultural beliefs have been appropriated by the pastoralists who still find value in the lake.

In addition to these religious beliefs, lakes also provide a space for recreational activities for both traditional village communities as well as other urban residents living around them. Walking, jogging and angling are common recreational activities to be seen around lakes. Villagers also gather around community spaces built around lakes to interact with each other over the day.

Modes of exclusion from access to and use of these lakes

All these uses of lakes notwithstanding, contemporary planning measures either tend to deliberately or unwittingly exclude traditional communities from using the lakes. Most lake rejuvenation and planning activities tend to be focused on creating enclosures, aesthetically improving the water body and providing recreational facilities in the form of neatly manicured lawns, boating, slides, swings, jogging tracks and exercising areas despite the fact that lakes in the city have not originated as or even transformed into mere aesthetic adornments to the landscape. Most such efforts at landscaping the lake result in the erection of fences and the enforcement of rules regarding who can access the lake, for what purpose and when. Consequently, while recreationalists get easy access to the lake, traditional users are excluded from the water body. This exclusion works in three ways. First, the physical boundary in the form of the fence serves as the primary deterrent to these users. Second, in order to enforce the entry restrictions, home guards are often deployed to guard the boundaries of lakes. These people prioritise certain users over others (such as the urban recreationalists over pastoralists) in keeping with dominant perceptions regarding the aesthetic and recreational utility of the lakes, providing the next level of exclusion. Thirdly, the imposition of entry timings around lakes (which are mostly around early mornings and early evenings) serve to isolate traditional users (mostly women) who come to collect fodder after their daily chores around the household.

These practices have meant that communities who once regarded the lake as their own now get distanced from the water body and adopt a stance of indifference towards its social-ecological condition. As a villager in one of the study sites in the heart of the city put it,

> They construct fences to keep us and our animals out. In their eyes, we are not good enough to enter the lake. Why should we go towards it then? Why should we care about what happens to it? It is of no more use to us. We don't care anymore.
>
> (Translated by the authors from the regional language of Kannada in which the interviews were conducted)

Policies such as privatisation of lakes that impose an entry fee on the user also serve to alienate marginalised lake dependent user groups from the resource. The new proposed smart city framework too makes little mention of the other forms of ecosystem services derived from lakes and green spaces outside of their recreational and aesthetic values. Regulation of traditional occupations (such as fishing) through tender-based processes has also served to isolate communities from a formerly important resource. The consequences of such modes of exclusion are discussed in the next section of this chapter.

Discussion

It is clear from the results presented in this chapter that the hegemonic neoliberal discourse has extended its pervasive reach into the governance and management of urban commons. Lakes are indeed dynamic spaces providing a wide variety of ecosystem services to an equally diverse set of users. We have also observed that traditionally, government or community led rejuvenation of lakes within the city of Bengaluru tends to prioritise dominant urban perceptions of lakes as aesthetic and recreational lungspaces of the city. In the process, it alienates a considerable number and diversity of users from an ecological common once integral to their way of life. This is rather ironic because these rejuvenation policies typically rely on ecosystem stewardship to maintain the rejuvenated water body. Ecosystem stewardship as the name suggests is possible only when communities relying on the lake are part of both the decision making and maintenance of the resource. In this case, ecosystem stewardship of rejuvenated lakes rests largely on an urban (often, migrant) population who have no real connection with the lake except for it being a pleasant diversion from the humdrum of everyday city life.

This exclusion of communities creates another set of problems. The alienation of people from formerly important resources reduces the perceived value of that resource in the minds of former users. As many of these communities are also marginalised and lacking in basic amenities such as toilets, lakes then begin to be used as sites for open defecation or for channelling raw sewage from these settlements, adding to the rather chaotic situation around them. In addition, the loss of perceived value of these water bodies can render them vulnerable to other threats in the form of increased pollution, encroachments, and conversion into built up structures (Unnikrishnan and Nagendra 2014). In fact many lakes of the city today are either severely polluted (the frothy lakes making headlines today are a case in point) or converted into malls, bus stands and sports stadiums. Tracing the history of these transformations reveals that these were largely due to the exclusion of those communities from the resource for whom it was an integral part of lives and livelihoods (Unnikrishnan and Nagendra 2016).

In addition, the character of lakes as ecological urban commons is severely compromised by such policies that enforce exclusion of people from a common pool resource. In the context of urban commons, this neoliberal regime ensures that the marginalised remain as such, while at the same time protecting the interests of sections of urban society who have the capacity to pay or perceive the resource in a certain manner. While studies of this nature are extremely limited and need additional evidence brought forward from other cities, the prioritisation of certain forms of ecosystem services over others as seen in this particular instance really brings us to the question: Whose commons is it anyway? On a broader level, it brings us full circle to the question of whose interest neoliberalism actually protects (Harvey 2007; Castree 2010). The answer to that may lead us towards progress in the direction of incorporating more inclusivity and justice in policy and planning around these resources.

Acknowledgements

The authors acknowledge early funding from a USAID peer grant to Ashoka Trust for Research in Ecology and the Environment (ATREE), Bengaluru, and research funding from Azim Premji University. A Newton International Fellowship awarded to Hita Unnikrishnan by the British Academy too is gratefully acknowledged by the authors.

References

Agarwal, A. and Narain, S. 1997. *Dying wisdom: Rise, fall, and potential of India's traditional water harvesting systems.* Center for Science and Environment, New Delhi.

Annaswamy, T.V. 2003. *Bengaluru to Bangalore: Urban history of Bangalore from the pre-historic period to the end of the 18th century*. Vengadam Publications, Bangalore.

Berkes, F., Feeny, D., Mc Cay, B.J., and Acheson, J.M. 1989. The benefits of the commons. *Nature* 340.

Buchanan, F. 1807. *A journey from the countries of Mysore, Canara, and Malabar*, vol. 1. Reprinted in 1999. Asian Educational Services, New Delhi.

Castree, N. 2010. Neoliberalism and the biophysical environment 2: Theorizing the neoliberalisation of nature. *Geography Compass* 4(12): 1734–1746.

Dikshit, G.S., Kuppuswamy, G.R., and Mohan, S.K. 1993. *Tank irrigation in Karnataka*. Gandhi Sahitya Sangha, Bangalore.

Harvey, D. 2007. Neoliberalism as creative destruction. *The Annals of the American Academy of Political and Social Science* 610: 21–44.

Kiran, R. and Ramachandra, T.V. 1999. Status of wetlands in Bangalore and its conservation aspects. *ENVIS Journal of Human Settlements* March: 16–24.

Mundoli, S., Manjunatha, B., and Nagendra, H. 2014. Effects of urbanisation on the use of lakes as commons in the peri-urban interface of Bengaluru, India. *International Journal of Urban Sustainable Development*. doi:10.1080/19463138.2014.982124.

Nair, J. 2005. *The promise of the metropolis: Bangalore's twentieth century*. Oxford University Press, New Delhi.

Narain, V. and Vij, S. 2015. Where have all the commons gone? *Geoforum* 68: 21–24.

Ostrom, E. 1990. *Governing the commons: The evolution of institutions for collective action*. Cambridge University Press, Cambridge, UK.

Ostrom, E. 2007. Institutional rational choice: An assessment of the institutional analysis and development framework. In *Theories of the policy process*, 2nd ed., P.A. Sabatier (ed.). Westview Press, Cambridge, MA.

Prudham, S. 2004. Poisoning the well: Neoliberalism and the contamination of municipal water in Walkerton, Ontario. *Geoforum* 35(3): 343–359.

Rice, B.L. 1905. *Epigraphia carnatica volume IX: Inscriptions in the Bangalore district*. Mysore Government Central Press, Bangalore.

Sudhira, H.S., Ramachandra, T.V., and Subrahmanya, M.H.B. 2007. Bangalore. *Cities* 24(5): 379–390.

Unnikrishnan, H. and Nagendra, H. 2014. Privatizing the commons: Impact on ecosystem services in Bangalore's lakes. *Urban Ecosystems*: 1–2.

Unnikrishnan, H. and Nagendra, H. 2016. Contested urban commons: Mapping the transition of a lake to a sports stadium in Bangalore. *International Journal of the Commons* 10(1): 265–293.

Part III

MARKETISATION AND THE ENVIRONMENT

8

PLAYING WITH COLOURED SPECTACLES

Neoliberal witchcraft as played out through watershed policies

Arun de Souza[1]

As children, many of us may have played with coloured spectacles. Putting them on was great fun – they gave us the illusion of seeing our familiar landscapes through a new hue. We knew the visions we saw were imaginary and that the game was in our control, it was pure fun. Watershed policy discourses are similar, but have a dangerous elision. We are presented a growing vision of democratisation, which hides the neoliberal paradigm[2] that undergirds it. Unlike the control we had as children over what colour we chose to use, this discourse is opaque and does not easily allow us to perceive how it tints. This chapter interrogates the Indian government's watershed development policy in order to reveal its neoliberal undergirdings that turn nature into a resource and human livelihood into an extractive and oppressive capitalist project. Watershed development projects are not just about conserving nature and allowing for sustainable livelihoods, they are actually projects that transform its participants into neoliberal subjects. This is not to say that the human subjects of this discourse are entirely powerless; through the micropolitics of local actors, these blueprints from a discursive centre are in small ways forged into either instruments of furthering the control of elites or subverted by the insurgency of marginalised populations.

Statist interventions to transform nature

Precolonial empires, like the Mughals, undertook agrarian land mapping exercises for the sake of revenue generation. But it was the colonial empire that engaged in large-scale land survey and settlement exercises on an all-India basis. They also engaged in large-scale resettlement and creation of agrarian production units in different parts of India. In the Indian subcontinent, "over the years between 1890 and 1970, more than thirty million hectares of land were transformed from forest and grassland into areas of crop production and settlement; the amount of land being cultivated rose by over 45 percent" (Tucker 1988: 119). In areas where the British established irrigation facilities, like the Punjab, agrarian resettlement communities were established (Gilmartin 1996: 217). Forests too began to be regulated for timber and other extracts. This reached its apogee in the enactment of the Indian Forest Act, 1865 (Guha 2000: 38, see also Grove 1998). This Act did not just elide the fact that forests were not free, unencumbered spaces, it transmogrified forest dwellers into encroachers on their own landholdings! Actually local tribal communities had used forests for their subsistence, medical, worship and aesthetic needs (Guha 2000: 38–43). For the administrators of the British empire, land, water and forests were commodities that needed to be put to rational use (Gilmartin 1996: 210–211, see also Deb 2009). In doing this, the British engaged in a simplification which failed to see the complex nested system of land rights and usages (Kotani 2002: 2–51).

Two of the important resources for the increase of foodgrain and cash crop production were water and land. The British began a series of massive interventions to transform this relationship between water and land in order to increase production. In the Orissa delta for instance the British began to build up a series of flood embankments and canals to increase the production of rice and cash crops. Through changes in the flood embankment systems the entire area's productivity relations were changed. This culminated in the post-independence Indian government setting up the Damodar Valley Corporation, modelled on the American Tennessee Valley Authority. Brute nature was converted into a "physical unit of the economy, a technical relation and a political artefact of capital" (D'Souza 2006: 239). Irrigation systems were also renovated or newly constructed all across India. Along with this soil and water conservation were undertaken in the United Provinces, Bombay and Madras Presidencies. The post-independence Indian government, with its ideas of socialist planning, continued this process of building large-scale irrigation works and dams (deSouza 2010: 20–21).

By the 1980s the government's attention began to shift towards dryland agriculture as it realised that 70 per cent of India's agrarian production came from this sector. Studies had also begun to point out that the government's soil and water conservation efforts were bearing very little fruit, mainly due to the piecemeal and ad hoc manner in which different government departments were carrying out these works. The Indian Council of Agricultural Research in 1982 began 47 model watershed development projects to integrate these efforts. In 1983 the World Bank stepped in with aid for some watershed development projects. The World Bank went on to fund the first Integrated Watershed Development Project in 1991. NGO participation brought in a new rhetoric of participatory development which the Hanumantha Rao Committee Report adopted. These were incorporated into the first watershed guidelines of 1995 issued by the Ministry of Rural Development. These guidelines were revised in 2001.[3] The 73rd Amendment inaugurated another slogan, that of development through the Panchayat. Its effects were seen in the Hariyali Guidelines of 2003 which encouraged Panchayats to take on the implementation of watershed projects. By 2006 liberalisation was becoming the new mantra, and based on the Parthasarthy report another set of guidelines was issued in 2008.

The neoliberal apparatus

One must remember that the 1980s, when the World Bank began to finance Watershed Development Projects, was also the time it began the Structural Adjustments Programme. Loans were being rescheduled and strict neoliberal policies began to be enforced. The World Bank's efforts were to institute a strong capitalist system of accumulation, to lessen the role of the state through deregulation and to ensure the rights of private property (see Harvey 2005). It is with this lens that one must view the discovery and consequent proliferation of watershed development projects.

Watershed development as witchcraft

The 1995 guidelines blandly state that the aim of watershed development is "to promote the economic development of the village community" (GOI 1995). This simplistic formulation sees human flourishing in only economic terms and papers over all the class, ethnic and power disparities within the village. In constituting the village as a community, the document unproblematically sees the interests and well-being of all members of the village as being similar. Later on it does seek to correct this by saying that watershed development should pay "special emphasis to improve the economic and social condition of the resource-poor and the disadvantaged sections of the Watershed Community such as the asset-less and the women" (GOI 1995). It goes on to say that this can be done through income generation activities and the setting up of 'self-help groups' (SHGs) for the poor. It even goes on to suggest the setting up of user groups in forest areas through the Joint Forestry Management scheme. This could have benefited the poor who use forest usufructs the most. The document recognises the difficulties involved in setting up user groups: The implementing agency "must recognise that people have different economic and social interests and they need to reconcile these differences to cooperate and work together for the holistic development of the watershed and at the same time ensure that the benefits are shared equitably with a tilt in favour of the poor and the weak, particularly when public investments are being made for the creation of common assets" (GOI 1995). Strangely in the 2001 guidelines this difficulty is elided and the document says user groups "shall be homogenous groups, who may be most affected by each work/activity and shall include the persons having land holding within the watershed areas" (GOI 2001). This formulation gets repeated near verbatim in the 2003, 2008 and 2011 guidelines (GOI 2003, 2008, 2011). In other words, user groups are now seen to be the preserve of the landed. Here lies the crux of the inequality inherent in all watershed development projects. While there is a recognition that there may be different categories of people in village society (caste, class and gender are recognised by the guidelines), there is no real attempt to address the concerns of those marginalised except for the hope of some sort of trickle-down effect through labour and SHGs.

Even worse, quite often watershed development projects in fact decrease the access of the landless to land resources and livelihood opportunities. In the watershed development project I studied (deSouza 2010), prior to the project a lot of the landless tribals in the village cultivated forest land for which they had recognised customary rights that had been recorded in the *Talathis* (village-level land records officer) office. The watershed development project necessitated these forest lands to be 'treated'[4] through contour bunding and reafforestation. As a result these landless were now evicted from their customary lands! Most of these landless also engaged in goat rearing that required them to graze on forest lands. This too was repeatedly curbed to prevent damage to young saplings. An experiment to train them to keep stall-fed goats failed as it was too cumbersome for them to bring in fodder everyday and disease easily spread amongst these closely kept goats. After some time the tribals began to take their cattle up the hillsides again, but this now became an uncertain venture as they could be impounded for trespassing on their own customary landholdings! What the British did through the Forest Act of 1865, the Indian government and complicit local elites are redoing to local landless communities: transmogrifying them into encroachers, trespassers on their own customary landholdings – neoliberal witchcraft at its best![5]

The magic of science and technology

Gyan Prakash (1999) in his magisterial study, *Another Reason: Science and the Imagination of Modern India*, shows how science was invoked and practised by the colonial British government as part of its 'civilising mission' as a way to improve and develop India. Science was seen as the way to rid India of its mythic thought and ushering in concrete improvements like the railways, irrigation, hydroelectric projects, public health and bureaucracy. Science in India has become both 'culture and power' (1999: 8). Prakash cites Kipling's *The Bridge Builders*, which narrates the story of the construction of a railway bridge over the Ganges, as a way to show how the British tamed India's gods and its people through the taming of nature (cited in

Prakash 1999: 5, 166–168). Nehru adopted this vision wholeheartedly and married it to the socialist rhetoric of science as signifying freedom. The right wing also adopted this understanding by tracing science's roots to India's ancient scriptures. Science thus became not just a way of knowing or taming nature, but also the way the state governmentalised India (1999: 10, 161) and constituted it as a developmentalist state.

One sees continuities of this developmentalist state in the way science, development and governance are sought to be linked together through the progressive importance given to the role of science and data in the policy documents. The 1995 document states, "Scientists have developed appropriate technologies" that are "unacceptable to the villagers on account of the socio-economic realities at the ground level which hinder its adoption" (GOI 1995). While it does recognise local knowledges that can be utilised in some places, it disdainfully sees local-level technologies as not being applicable for the watershed as a whole. Village people "need training and exposure to modern scientific and technical methods, entrepreneurial skills to identify and exploit opportunities" (GOI 1995). Similar ideas are repeated in the later documents. Policy makers seem to have disregarded centuries of experimentation in conserving and transporting water for agricultural and human needs in favour of what science has to offer in terms of higher outputs and neoliberal entrepreneurship skills.

While the 2001 policy merely states that remote sensing data should be utilised and institutions of recognised excellence in science should be consulted, by the 2008 document this reaches a new height with the suggestion to establish a National Data Centre, National Portal and a state-level Data Cell. The 2003 policy advocates the setting up of these data centres at the district level too. The 2003 policy has this grandiose desire: "The National Data Centre (NDC) would collate summary data for the entire country, archival data, data for programme and fund flow management. This centre is planned to be equipped with various GIS[6] thematic layers for cadastral, watershed, soil, land use, socioeconomic parameters, habitation etc." (GOI 2003). While a lot of this remains wishful thinking, today ground-level NGOs are able to access basic GIS data and know how to prepare glossy 'detailed project reports' (DPR) (required by the government to pass a project) and flyers with this colourful data presented in thematic layers. Government-appointed teams are exhorted by this document to continuously monitor these projects and feed information into these data centres. Today at the click of a mouse, the National Rainfed Area Authority (NRAA) can view ongoing work and its status even if the village is remote. Satellite imagery is available for every part of India.

British technology was not as well developed as today's technology. Surveillance and control over state bureaucrats and its citizens is much easier today. One must also remember that these dryland areas were precisely the spaces over which the state had previously very little control or oversight.

Through watershed development projects and the related monitoring technologies the penetration of the state and its surveillance has acquired a major fillip. This does not make the state more efficient in its delivery of welfare services as James Ferguson (1994) so trenchantly pointed out. What it does is make for a system of thought where everything is seen as quantifiable and categorisable. The magic of science creates citizens who become adept at citing data and makes state categories real. As the project begins to unfold people learn new technologies and terminologies. Words like contours, gully plugs, check dams and cubic feet of water become common parlance. Western science and private enterprise becomes a way of life. They begin to valorise private enterprise over communitarian practices. Just as 'scheduled caste' today is not just a recognised bureaucratic category but also an identity marker, in the same way, land, water, trees, crops and human beings become economic categories through which people view progress. The village I studied was near Ralegaon Siddhi, the iconic village in Maharashtra for watershed development. All kinds of indicators of comparison would be thrown at me – see how many buildings Ralegaon has, see how much water it consumes (some would even quote statistics to show this), see how its crop output has increased. Development and personal identities become quantifiable and nameable through a set of numbers that also get linked to identity and status. Ralegaon Siddhi or Hivre Bazar are today not just places in which people reside, but get talked about throughout the global circuit of development practitioners and aid agencies. They become iconic figures for what development means, for what citizens must aspire to.

The panacea of scientific management

Leslie Sklair (2015) links globalisation to a growing class of transnational elites. For these elites the corporate sector and its business practices are the template for all other forms of governance. Benchmarking through global best practices, the setting in place of clear hierarchical systems of authority that are led by chief executive officers (CEOs) of calibre, the setting of verifiable targets and flexibility of practice are some of the important characteristics of this form of governance that is sought to be valorised over bureaucratic, corrupt and slow-moving state practices. Sklair speaks of top-level corporate executives, bureaucrats and globalising professionals as constituting this class of transnational elites (see also Mosse 2013: 225).

Top-level leadership of institutions like the World Bank, bureaucrats and executives of NGOs are all a part of this class or are at least exposed to the ideas of this class. It is this class that has slowly transformed the watershed guidelines' language and the institutional procedures set out therein. The 1995 guidelines spoke of the "inadequate management skills" of the project staff and villagers. An NGO is the preferred "project implementation agency" (GOI 1995). There is no mention of any specialised supra village level entity at this point of time. The 2001 guidelines spoke of the preparation of a "perspective plan" at the village level. By the 2008 guidelines an entire chain of command is established. The 2008 guidelines mention the NRAA as the nodal agency for the pan-India watershed programme, with each ministry having its own nodal agency. The NRAA is to set up a National Data Centre and National Portal. A state-level nodal agency (SLNA) with a CEO (note the terminology!) chosen from within the bureaucracy or a professional from outside is appointed. Below this comes the District Watershed Development Unit (DWDU) with its own data centre. One can see the proliferation of agencies that are slowly sought to be taken out of the ambit of government control (i.e., its political arm) and handed over to a technocratic system of governmentality. The NRAA is an appointed nodal agency and the ministries are tolerated as having their own supervisory role, but the document's own rhetoric would seem to have preferred the ministries being taken out of the game with the NRAA as having its own CEO.[7] Monitoring and constant data feedback is encouraged through the use of latest satellite and internet technologies. Documentation of 'best practices' is a necessary part of this feedback loop.

In his study of tank irrigation projects in South India, Mosse points to a similar process of converting local level 'traditional' subjects into modern subjects of the global corporate discourse:

> sociologies of village irrigation tradition are, in a sense, shaped by particular forms of government, it may not be surprising to see, under the global (and especially from 1991, Indian) re-emphasis on the free market and private management, a reformulation of old communal ideas within the irrigation policy community. In this externally driven ideological reorientation, farmers (or *ryots* – colonial tenants/subjects) become 'users' or 'appropriators', their customs and traditions become 'social or institutional capital,' and village organisations become 'water management associations'.
> (1999: 324)

This transformation of terminology for watershed development institutions is not just a renaming of what already exists. One, this terminology robs local institutions and functionaries of all other interests except the economic. A group of individuals who constitute a Village Watershed Committee (VWC) are seen to be rational maximisers whose sole interest lies in conserving and provisioning water in an efficient way. All other interests (e.g., political, symbolic, cultural as we shall see later) are seen as outside this rational managerial framework. Further, it magically waves away the entire history of water usage and resource within the village (Mosse 1999: 326). This is a history wherein land was inequitably distributed and many were left landless. Sociological deconstruction is necessary to bring this inequitable history back into focus.

But most importantly, this terminology and its associated practices, e.g. minute books and accounts registers, create a new subjectivity. This subjectivity begins to pervade the practices of the village as it begins to set up other institutions with similar rationales and practices. As Mosse points out, "Labels serve to 'stabilize the flux of life'; but not only this, they also change behaviour and create (or eliminate) the realities they identify" (1999: 331–332). Any new organisation set up in the village tends to imitate these new practices. They no longer become impositions from above, but the valorised way of normalised practice. Corporate sector management practices become valorised over traditional modes of cooperation which begin to be subtly looked down upon. Individualisation of the subject becomes a natural consequence. I once saw a tragic example of this: When a family lost its earning member, neighbours advised the wife to approach the local credit cooperative for a loan. As an elder pointed out to me, in a previous era, it would have been natural for the neighbours to pitch in with this required help.

Subjects of a neoliberal discourse

Through the aforementioned processes, people residing in these far-flung regions that had previously been outside regular state control are now transformed into disciplined neoliberal citizens. We need to become aware of what "development conceals – especially strategies of power. Development agencies' claims to improve the conditions of other people disguised governmental practices of control and incorporation of 'dangerous borderlands' into the state's grids or global capitalism" (Mosse 2013: 229). Most of these villages, one must remember, are dryland villages, usually far from the normal wetland irrigation systems that the state normally concentrated on. Village peoples, who were otherwise far from state practices, inexorably become citizens whose main and only aim is to increase their livelihood (read: earnings that can be used to consume the goods produced by neoliberalism's factories). Evaluations of the success of watershed development programmes invariably involve the measurement of increased consumerism (e.g. the buying of TVs, motorcycles, fridges). They gain an economic citizenship that devalues all their previously variegated cultural values. This cultural paradigm is one in which Western scientific rationality and its categories of thought become their cultural paradigm, a paradigm that gains a ferocious power over them. This power is maintained and practised through the capillary processes of modern managerial techniques that are heavily dependent on the new information and communications technologies (ICTs).

This process of converting them into neoliberal economic citizens is masked through the varied hues that tint the document. In general, each guideline is supposed to have increased the process of democratisation. The 1995 guidelines were celebrated for the way they allowed for the participation of the stakeholders. Later, when the 73rd Amendment involving panchayati raj institutions began to proliferate, many activists, politicians and bureaucrats got carried away by this rhetoric. Thus the 2001 guidelines put the Panchayat as the linchpin of the programme. The 2008 guidelines even-handedly allow for the Panchayat or an NGO to become the project implementing agency. The 2008 guidelines based on the euphoria associated with new ICTs claim to inaugurate a new era of further democratisation through technology that will allow for constant feedback and monitoring. In its technology section the guidelines sing the paeans of technology and its relationship to a new India: "Technology enables us . . . [to] *facilitate the free and seamless flow* of information and data" (emphasis added by me). "Thus, the endeavour would be to build in strong technology inputs into the new vision of watershed programmes. . . . Thus technology inputs would bring about a paradigm shift in the implementation and management of the area development programmes" (GOI 2008). Freedom and development get inextricably linked to the spread and use of ICTs. The 2011 guidelines further this process of seeing ICTs as the panacea for rural transformation. However, as the previous sections have sought to show, even as the 2008 and 2011 guidelines speak of equity, gender equality, decentralisation, inclusive growth, local capacity building, community participation and facilitation, actually in effect, behind the scenes, a new neoliberal global citizen is being forged. Some of these citizens are given more powers, others are pushed into the disempowered labour class, the fodder required to run neoliberal enterprises.

Contestation, subversion and co-optation

While the state and its agencies try to impose a neoliberal discourse upon citizens, at the grassroots other countervailing processes do take place. Watershed projects are not just sites for the production of neoliberal citizens given to the virtues of economic profit. Citizens engage in processes that also cater to their varied needs. These needs could lead to the solidification of older hierarchies or result in the challenging of these hierarchies.

Establishing Maratha dominance

The dominant caste of the watershed project that I studied was made up of the landed Maratha community. The major political parties were the Nationalist Congress Party (NCP) led by Sharad Pawar and the Bhartiya Janata Party (BJP). There were two important political institutions through which these parties sought to establish control over the village. These were the *Vividh Karyakari Sanstha* (VKS, i.e., Multipurpose Society) which channelised most of the government projects and schemes to the village and the

Panchayat. Elections to both these bodies were hotly contested. The VKS elections was the most important battleground as the funds here were large and were constantly replenished through the varied schemes of the government. The NCP controlled the Panchayat, while the BJP controlled the VKS.[8] However, both these parties were dominated by the Maratha community with a token percentage of other castes. As far as the scheduled castes and tribes were concerned, these were only represented on the statutorily required reserved seats (usually about 1–3 members).

The VWC, which was the statutory body supervising the Watershed Project, was supposed to be independent of the politics of the village and hence had no elections based on party affiliation. Members were supposed to be elected by a majority vote of the gram sabha, i.e., the people involved in the watershed project, in theory, men and women of all castes within the catchment of the watershed. Whenever I sought to find out the exact date and time for this election, I would be fed with different venues, dates and different sets of names. It was only much later that I realised what had happened. The NGO concerned had engaged in a delicate negotiation with all its major power wielders. Like the usual scenario in which a cabinet is selected, each of the major parties were represented on the VWC through their proxies. The president and the secretary came from the BJP, while the vice president and the treasurer were affiliated to the NCP. Any major decision would involve a negotiation between these four persons. Usually the other members (especially women, Scheduled Caste (SC) and Scheduled Tribe (ST) representative) then signed on the minute book. Meetings of all members were very rarely held, and these too mostly on the occasion of visits by the funding agencies.

It is also interesting to note the spaces within which a lot of these decisions get debated and negotiated. On two different occasions I watched how contracts for work like digging or desilting were handled. In one case, the person who owned the tractor which had done the ground levelling work was pressing for his payment from the VWC president. The president, sitting in the office of the BJP-controlled credit cooperative, called up the treasurer who was affiliated to the NCP and whose party also controlled another local credit society in the village. The treasurer came over to the BJP-controlled credit society's office where this negotiation was taking place. He placed on record that the man concerned owed the NCP affiliated credit society a certain sum of money. They then hammered out a deal wherein some of the money paid to the contractor would be immediately deposited into the NCP's credit society, a certain percentage would be kept on hold by the president as a 'security deposit' for the contractor's ongoing work, and the rest paid to the contractor. The results of this long, drawn-out negotiation satisfied the needs of all – the NCP person had ingratiated himself to his local political bosses who had got back some of their loan, the president, under the guise of a security deposit, gained some money to play around with, and the contractor too got his share. While I have no proof, there were also allegations that the money being calculated in favour of the contractor was more than the actual work done by him. One can see that while each of these political parties attack each other at election time, they usually on the day-to-day level engage in subtle negotiations that benefit both sides. In fact, at one of my interviews this became quite clear. The Panchayat elections had taken place sometime prior to the VKS elections. While interviewing one of the NCP local *pudharis* (marathi for 'leader'), I asked him what were the accusations against the opposition that lost them the Panchayat elections. He said they had engaged in large-scale corruption. I later on moved to the VKS elections where the NCP had lost and asked what were the charges brought against the NCP that resulted in their loss. He paused a moment, smiled, and said in marathi, "*Te´ch*," i.e., "The same!" In other words, the accusations they make against each other are usually convenient mirror images! The more important consideration is that Maratha dominance over village society continues no matter which party wins. It is this same pattern that was repeated, albeit through the NGO's negotiations, in the control over the VWC.

Similar processes are at work in the construction of structures to retain water and allow it to percolate into the soil. Check dams are built along the rivulet in order to slow down the speed of the water. This serves both to decrease soil erosion and to allow the standing water (behind the check dam) to percolate into the soil thus recharging the groundwater aquifers. This usually results in better water output for nearby wells. In theory, the process of choosing where to site a check dam is a technical process. Engineers are supposed to assess where best to situate a check dam based on soil structure and the suitability of the rivulet's embankments. In practice, politically powerful landowners work on the NGO representatives to build these check dams upstream of their landholdings so that their own wells and landholdings benefit. Quite a few such check dams were sited in places that are convenient to the landowner rather than at the technically best suitable place. Usually such check dams are recognizable through the colloquial name they acquire. Rather than using the technical number of the check dam that exists on the village-level watershed map, these are named something like, *maji sarpanchche* ('ex-Sarpanch', the village Panchayat's chief functionary) check dam. This means that this check dam serves the interests of the previous Sarpanch best. When the NGO representative was queried about this by me, his answer was: "There were actually two or three suitable sites. We chose this one because we knew that cooperation was assured. The *maji sarpanch* got his family members to clear the area of undergrowth, he provided the required black soil and he constantly supervised the work to ensure its quality. It's a win-win situation on all sides." The rhetoric of efficiency, cooperation and quality work hid the politics of power that underlay this construction.

Insurgency at the margins

On the other hand, those on the margins of the project do not find any direct benefit from the project as most of its gains, through increased water-tables, go to the landowning class. There are supposed to be trickle-down benefits – higher wage earnings due to increased cropping cycles and better fodder for those involved in goat rearing. In practice these trickle-down benefits are of a doubtful nature. Increased earnings only come during specific seasonal cycles leaving the landless without work in out-of-season times like the summer. They also find that goat rearing becomes difficult as the jungle and the areas they would take their goats to become fenced in due to reafforestation. As we saw previously, attempts to change their livelihood patterns by shifting to stall-fed goat rearing was unsuccessful.

In order to cope, they sometimes voted with their feet. During the May season, when demand for watershed labour was the highest, the Thakkars (a tribal landless group) went off to the Konkan belt for labour. The Marathas were furious and told me how stupid the tribals were. '*Tyana hindnyachi hous ahe*': 'They love to roam around' was their refrain. The Thakkars however had a different story. When the watershed was not around they had no labour in the nearby area nor did any Maratha family help them. They had thus built up contractual relationships with horticultural landowners in the Konkan belt. Most often they were involved in tamarind plucking and drying work or on construction labour. These landowners would pay them an advance for the following year too and this would help the Thakkars to tide over the lean season and also ensure for the Konkan landowner that they would come back the next year. This resulted in the NGO having to import labour from outside for seasonal work.

At seasonal times when the Marathas required the Thakkars for labour, the Marathas would come over to the Thakkar hamlet and find the youth playing cards early in the morning. They would then cajole them and promise higher wages in order to get them to work. The Marathas would later tell me: "See how lazy these people are. They can afford to play cards while we are desperate about the condition of our farms. We have to act nice to them to get them to work." The Thakkars on the other hand were quite clear that these same Marathas would previously take them for granted. Now that they required the Thakkar's labour, these people would play around to ensure higher wages. Also, for the Thakkars, the question of working hard on their landholdings did not arise, as they owned very little land. They thus rejected the label of laziness.

Land use values are not always based on livelihood needs. This is something that the policy documents elide. For instance, the Thakkars had a plot of land in the forest where they buried their dead. This was unlike the practice of the Marathas who cremated their dead. At some point a few acres of forest land was given over by the government through a scheme for ex-servicemen to a person from the village. The NGO wanted to conduct afforestation work on this plot. The ex-serviceman was willing since this would result in horticultural plants which he could harvest. The Thakkars however objected as one plot within this was where their ancestors were buried. They subtly raised the issue of how the ghosts of their ancestors might haunt the project and themselves. This evoked the residual superstition of the ex-serviceman resulting in worry for his personal safety, and the NGO intervened to hammer out a compromise. The Thakkars would have access to their burial ground (in spite of it being owned by the ex-serviceman) and the NGO would plan fruit trees on the rest of the plot for a reduced rate. The transcripts (cf. Scott 1992) of death, ghosts and haunting shared by all in this community were thus manipulated to the benefit of the Thakkar community.

The Thakkars also learnt to engage in the micropolitics of exchange deals. At a point of time when their work was required desperately by the NGO, they pointed to the sad state of their houses as the reason for why they could not come to work. One of the retaining walls in their Indira Awas Yojana house colony had fallen down. The Marathas then worked on the NGO to demarcate some of their funds for the repair of this wall and work on the watershed continued smoothly.

The Thakkars had thus fine-tuned a strategy of subversion and accommodation. At times they boycotted the work, at other times they complied through a negotiated exchange deal. This is not to say that they had managed to overturn the power structure. However, their micro politics had ensured some minor benefits for themselves. One must not unnecessarily romanticise these deals as if the discourses of dominance had been weakened. This perhaps is part of the age-old weapons of the weak that Scott (1985, 1992) has so eloquently detailed. When faced with such a huge inequality in power, the weak resort to guerilla warfare in order to eke out minor ameliorations to their hugely oppressive social relationships. Most often, NGO representatives saw these subversions in terms of opposition to the project, as not working for the common good. Very rarely were they able to see these in terms of opposition to the inequality and oppression that these groups faced.

Conclusion

The policy guidelines around watershed development have quite clearly presumed the only value of water and land to be that which is economic and seen primarily in terms of livelihood supports. They have hidden this understanding through a pretence at increasing processes of democratisation. At some points these coloured lenses that blurred our vision were labelled as participatory governance or local-level panchayati raj, at other times newer technologies that were purportedly meant to increase freedom (ICTs) and efficiency (managerial practices) were valorised. In practice, most of the benefits went to the landed. The landless often lost out on their customary landholdings and were further divested of livelihood opportunities. The elite thus increased their wealth, while the poor faced further deprivation. Harvey aptly sums this up in his analysis of how neoliberalism acquires power:

> The main substantive achievement of neoliberalisation, however, has been to redistribute, rather than to generate, wealth and income. I have elsewhere provided an account of the main mechanisms whereby this was achieved under the rubric of 'accumulation by dispossession'. By this I mean the continuation and proliferation of accumulation practices which Marx had treated as 'primitive' or 'original' during the rise of capitalism. These include the commodification and privatisation of land and the forceful expulsion of peasant populations; conversion of various forms of property rights into exclusive private property rights; suppression of rights to the commons; commodification of labour power; and the suppression of alternative (indigenous) forms of production and consumption.
> (2005: 159)

It is not as if these policy paradigms have gone uncontested. On the ground people engage in their own local-level micro politics. Elites seek to capture these projects as ways to shore up or to increase their own control over the social system. On the other hand, marginalised groups seek to use the project to carve out miniscule spaces of survival from gross oppression. People seek to assert values that the policy documents may leave out: What for some may just be forest plots in need of afforestation, may be for marginal groups plots for cultivation, areas to graze their cattle on or places to bury their dead. Marginalised groups have to continually engage in a guerilla warfare to assert or protect their rights. Like Gramsci's war of position (2007), this intermittent guerilla war is part of the larger ongoing battle for marginalised groups to carve out their own spaces which the state seeks to crush and obliterate. Statist forces, elites and those on the margins are constantly engaged in this war of position with no clear winner; what takes place are only minor battles that shift the status quo marginally. Activists and NGOs need to widen these spaces of contestation consciously rather than condemning them as opposition to their Watershed Project plans.

Notes

1 A previous version of this chapter was originally published as Arun de Souza, 'The Coloured Lens Effect: Disentangling the Contradictory Discourses within Watershed Policies', *Sociological Bulletin*, Vol. 67, Issue 3, pp. 290–301. Copyright 2018 © Indian Sociological Society. All rights reserved. Reproduced with the permission of the copyright holders and the publishers, SAGE Publications India Pvt. Ltd, New Delhi.

2 I use neoliberalism in the sense that David Harvey defined it: "Neoliberalism is in the first instance a theory of political economic practices that proposes that human well-being can best be advanced by liberating individual entrepreneurial freedoms and skills within an institutional framework characterised by strong private property rights, free markets and free trade. The role of the state is to create and preserve an institutional framework appropriate to such practices" (2005: 2).

3 In 2000 the Ministry of Agriculture revised its own guidelines for the National Watershed Development Project for Rainfed Areas (NWDPRA). For the sake of convenience, we shall leave this document out of our analysis.

4 Note the language used by the government and NGO: 'treatment', as if to say that the tribals had infected the land negatively through their usage of it!

5 See the section titled 'Colonial Interventions', para 1.

6 Geographical information system based on satellite imagery. It contains details of topography, weather, soil types, cropping patterns and human population.

7 This chapter might give the impression that the 'State' is a monolithic institution. But this instance of two narratives of control sitting uncomfortably with each other is a good way to see how different arms of the same government may pull in different directions and how policy documents are an attempt to paper over these differences. In this instance 'technocratic managerial rationality' is pitted against the 'realpolitik' of the ministries concerned.

8 Legally, Panchayat elections had to be fought through locally constituted panels. While these gave themselves names like *Shetkari Manch* (i.e., 'farmers panel'), in practice these were affiliated to state-level parties like the BJP and NCP.

References

Deb, Debal (2009) *Beyond Developmentality: Constructing Inclusive Freedom and Sustainability*. London: Earthscan.

deSouza, Arun (2010) *Water and Development: Forging Green Communities for Watersheds*. Hardback. New Delhi: Orient BlackSwan.

D'Souza, Rohan (2006) *Drowned and Dammed: Colonial Capitalism and Flood Control in Eastern India*. New Delhi: Oxford University Press.

Ferguson, James (1994) *The Anti-Politics Machine: Development, Depoliticisation, and Bureaucratic Power in Lesotho*. Minnesota: University of Minnesota Press.

Gilmartin, David (1996) "Models of the Hydraulic Environment: Colonial Irrigation, State Power and Community in the Indus Basin," in *Nature, Culture, Imperialism: Essays on the Environmental History of South Asia*, David Arnold and Ramachandra Guha (eds.). New Delhi: Oxford University Press, 210–236.

Government of India (GOI) (1995) *Guidelines for Watershed Development Programme*. http://www.rainwaterharvesting.org/downloads/watershed.htm. Accessed on 2 Jan. 2000.

——— (2001) *Guidelines for Watershed Development (Revised – 2001)*. http://dolr.nic.in/dolr/guidewd.asp. Accessed on 1 Oct. 2015.

——— (2003) *Guidelines for Hariyali (2003)*. http://dolr.nic.in/dolr/Hariyali Guidelines.asp. Accessed on 1 Oct. 2015.

——— (2008) *Common Guidelines for Watershed Development Projects, 2008*. http://dolr.nic.in/dolr/downloads/pdfs/CommonGuidelines2008.pdf. Accessed on 1 Oct. 2015.

——— (2011) *Common Guidelines for Watershed Development Projects-2008*, Rev. ed. New Delhi: NRAA.

Gramsci, A. (2007) *Prison Notebooks*, Vol. 3. J.A. Buttigieg (trans.). New York: Columbia University Press.

Grove, Richard H. (1998) *Ecology, Climate and Empire: The Indian Legacy in Global Environmental History, 1400–1940*. New Delhi: Oxford University Press.

Guha, Ramachandra (2000) *Environmentalism: A Global History*. New Delhi: Oxford University Press.

Harvey, David (2005) *A Brief History of Neoliberalism*. Oxford: Oxford University Press.

Kotani, Hiroyuki (2002) *Western India in Historical Transition: Seventeenth to Early Twentieth Centuries*. New Delhi: Manohar Publishers and Distributors.

Mosse, David (1999) "Colonial and Contemporary Ideologies of 'Community Management': The Case of Tank Irrigation Development in South India," *Modern Asian Studies*, 33 (2): 303–338.

——— (2013) "The Anthropology of International Development," *Annual Review of Anthropology*, 42: 227–246.

Prakash, Gyan (1999) *Another Reason: Science and the Imagination of Modern India*. Princeton: Princeton University Press.

Scott, J.C. (1985) *Weapons of the Weak: Everyday Forms of Resistance*. New Haven and London: Yale University Press.

——— (1992) *Domination and the Arts of Resistance: Hidden Transcripts*. New Haven and London: Yale University Press.

Sklair, Leslie (2015) "The Transnational Capitalist Class and the Discourse of Globalisation," in *The Globalisation and Development Reader: Perspectives on Development and Global Change*, 2nd ed., J. Timmons Roberts, Amy Bellone Hite, and Nitsan Chorev (eds.). Oxford: Wiley Blackwell, 304–318.

Tucker, Richard P. (1988) "The Depletion of India's Forests Under British Imperialism: Planters, Foresters, and Peasants in Assam and Kerala," in *The Ends of the Earth: Perspectives on Modern Environmental History*, Donald Worster (ed.). Cambridge: Cambridge University Press, 118–140.

9

CANAL COMMANDS AND RISING INEQUITY

Seema Kulkarni[1]

Public irrigation system or canal irrigation has been central to the waterscapes of India's social and political history. Its expansion occurred towards the end of the 19th century when massive investments were made in large-scale irrigation projects which became a major source of revenue collection for the colonial empire.

From 1951, when the planning phase started in India, until about 1983, there was a steady increase in surface irrigation with an expansion of about 2.2 per cent annually. For the first time however in 1970–1974, the area irrigated by wells surpassed the area irrigated by canals. After 1975, the gap between the area irrigated by wells and that irrigated by canals was widening. The face of public irrigation system has been changing rapidly since then. However, four decades later, we are still grappling with policies related to the relationship between surface and groundwater and how best to manage and govern both. The invisibility of groundwater and the significant role of the visible surface water in the recharge of this invisible water have further compounded the question of governing this resource, thereby leading to various forms of inequities.

While the immediate post-Independence era focused on dams and development and symbolised dams as the temples of modern India, the increasing thirst for water for development based on intense use of fossil fuels (green revolution agriculture and urbanisation and industrialisation) prompted the state to heavily invest in groundwater development as well. The 1970s marked a watershed in this respect. The ease of access and the command and control a user/owner can have over the resource led to a huge expansion of groundwater in India. Presently, around 85–90 per cent of rural habitations in India depend entirely on groundwater for their domestic needs, and groundwater constitutes about 65 per cent of India's net irrigated area, indicative of India's dependence on groundwater (Kulkarni and Vijayshankar 2014).

The rush to exploit groundwater without any regulatory norms lasted for over two decades after which concerns around scarcity, depletion and poor water quality started hounding the state and the users. The growing thirst has literally led to the "race to the bottom" in terms of exploitation of groundwater.

While groundwater has been privately owned, the state played a major role in its development but little or no role in governing its management and use. In a separate review of the sector as a whole at the country level, Shah (2011) very succinctly points to the reality of an Indian irrigation system which he says never seems to conform to its design. The central problem is 'unauthorised' over-appropriation of water by head-reach farmers for growing crops that irrigation planners had never expected them to grow. He points to a complete non-recognition of the emergence of what he refers to as a pump irrigation economy in which scavenging water from any proximate source – ground or surface – has taken precedence over orderly gravity flow irrigation (ibid.).

The post-1990s has seen significant shifts in the policies around water and agriculture. Promoting subsidies in commodity crops and extending water rights to non-agricultural uses has led to several reallocations in water that has changed the land, labour and agrarian relations. Inequities have been on the rise and the 'scarce' resource is now getting concentrated in the hands of few. Knowing and acknowledging the role of irrigation in increasing agricultural production warrants a timely intervention both on the policy front as well as in practice to extend the outreach of irrigation to a large number of farms and people in order to support their livelihoods.

Maharashtra: overview of irrigation policies

Maharashtra is among the states with the highest number of large dams, but ironically with only a little over 17 per cent of cropped area irrigated against the all-India figure of about 45 per cent. Given that 83 per cent of the cropped area is unirrigated and 75 per cent of the area is classified as drought prone, expanding irrigation to larger number of farms and people acquires a lot of significance.

The state saw a lot of irrigation projects coming up in the late 19th century primarily as part of public works towards famine relief. Historically, Maharashtra had several community managed irrigation systems, such as the Phad[2] system of North Maharashtra or the Malgujari[3] tanks of Vidarbha, which were decentralised and practised some forms of equitable sharing of water and shared governance. These were of course located in the caste and patriarchal contexts and so did not allow for water sharing outside of the agrarian community. But these systems had several advantages over the more centralised systems that came in later during the British raj.

The late-19th-century infrastructure remained underutilised as the cropping patterns were such that they required very little water. Besides, the British imposed stringent rules about payment of water charges in advance

(Datye et al. 2004). Thus, utilisation of these large systems could not get the British the revenue they hoped to collect from water.

In 1938 under the leadership of Sir Visvesvaraya,[4] an irrigation enquiry committee was set up to look into the various problems faced by the water sector, especially the causes of the underutilisation of the canal systems. The committee recommended that the irrigable area under each canal should be distributed equitably by giving each village and as far as possible each cultivator just enough land so that he/she can manage irrigated crops properly. The suggested norm for allocation of the *bagait* or the irrigated land to be allotted to a village was one-third of the total cropped area of that village. As early as 1938, the committee recommended the formation of users groups and volumetric supply of water. It involved the farmers in decision making and planning of irrigation. It is said that Sir Visvesvaraya used to hold large meetings of farmers to get their viewpoints regarding crops and water rotations (Government of Maharashtra [GoM], 1999; Lele and Patil 1994). None of these recommendations were taken very seriously, so various alternatives were introduced to improve the uptake of irrigation. One of them was the block system in the Nira Pravara irrigation system. In fact, this laid the foundation of an inequitable system and the basis for concentration of water rights in the hands of few. The block system is a long-term sanction of water rights to a few farmers. This unequal system continues to date and has firmly established itself, making it difficult to change despite the new commitment to equity. The block system is indicative of the political economy of water, just as the cropping choices and shifts in policies that we see later.

With every passing year, irrigation became a vexed problem requiring serious attention. The government thus appointed a series of committees and two major commissions to look into the irrigation problem. In 1962, the first irrigation commission in fact clearly indicated that only 30 per cent of the cropped area will be irrigated with full potential of irrigation development. There was a recommendation that unless there is a ceiling on area irrigated by cash crops, the benefits of irrigation would not extend to larger populations. The Barve Commission, as it was known, also made it clear that development of groundwater should be in areas outside of the canal commands. However, none of these recommendations were taken seriously thereby aggravating the problem further.

The 1972 drought of Maharashtra put the question of water back on the map of the state. A fact-finding committee on drought in 1973 suggested that storage capacities of irrigation projects should be expanded and that protective irrigation for a large number of farms and people should be the norm for drought-prone areas. It lowered the stringent requirements of constructing dams at 75 per cent dependability to 50 per cent dependability thereby making it easier for irrigation projects to be sanctioned in drought-prone areas. This was followed by the famous 3-D Committee or the eight monthly irrigation committee of 1979 which had the three Ds, that is, Dandekar, Deuskar and Deshmukh,[5] three significant people who influenced the policies and politics of Maharashtra and proposed reforms to extend irrigation into the drought-prone areas. The committee proposed that one-third of the irrigation water should be used for kharif irrigation and two-third for rabi and hot weather crops. It also stated in clear terms that irrigation should not be extended to annual crops in water deficit regions of Maharashtra.

There was also a committee to look into the irrigation backlog of the Vidarbha region which proposed several measures to lift the region from its backlog. The second irrigation commission, also known as the Chitale Commission, was set up in 1999[6] and it has given a very comprehensive set of recommendations to improve the water sector in the state. As per the Chitale Commission, which has taken a more integrated view of water, it is possible to attain an ultimate irrigation potential to up to 62 per cent of the total cropped area. The commission looks at the water sector as a whole rather than focusing merely on surface systems. It calls for a river basin planning approach (GoM 1999). The present irrigation potential reached is only about 32 per cent of what has been projected by the Chitale Commission. However, it is still not clear as to whether the Chitale Commission's recommendations are ratified by the government. None of them seem to be applied.

To conclude this discussion, irrigation reforms in Maharashtra until the late 1990s focused on increasing the irrigation potential through large-scale investments in irrigation infrastructure. It appointed several committees periodically to look into the crisis within the sector. However, until 1999 the committees appointed did not look beyond the surface systems, perhaps because groundwater had not pervaded the canal commands until then. This myopic view led to policy responses that ignored the burgeoning groundwater sector and its increasing interface with the canal commands. But it is also important to note here that all the committees were largely speaking a language that curtailed the unabashed use of water for annual crops and that left the drought prone areas dry and its people poor.

The conventional approach to govern surface water management has failed miserably in Maharashtra and many other states in India. Across India, we see a huge performance gap in the created irrigation potential and the actual irrigated area. In Maharashtra, for example, the actual irrigated area is less than 50 per cent of the created potential (Purandare 2012). There have been no ground-level studies to understand why this has happened. Without such studies, the recommendations are often directed towards infrastructure improvement rather than looking at and responding to the underlying causes that affect underutilisation of created potential. So, it is often good money going for bad (Shah 2011) and a business-as-usual scenario where less and less is invested to understand the rapid changes that the public irrigation sector has seen over the last few decades and more and more for infrastructure repair.

Reforms in the last two decades (2000–2017)[7]

Since 2000, with small beginnings in the mid-nineties, the state was clearly moving towards a more market-based approach to water. The Dublin conference held in 1991 is considered to be a milestone in defining and framing water as a commodity rather than as a social good. Although there have been debates around whether or not water is a social or an economic good or has elements of both, the central point is that the Dublin principles marked the beginning of integrating water into the market system. The process has varied across the globe. In India, withdrawal of the state from being a provider to being a facilitator is evident from the various policies it introduced in the water sector. The Swajaldhara programme for drinking water is an example of that. Community participation was invoked in contributing to partial capital costs towards infrastructure and the entire management of operation and maintenance. The framing of the water problem shifted from lack of infrastructure to lack of governance and scarcity, thereby emphasising the need for institutional and economic reform. So, although the state showed withdrawal in provisioning, it played a proactive role in setting up policies that ensured that the 'scarce' water resource is governed better through appropriate institutions and pricing mechanisms.

While irrigation policies and priorities were changing, so were the agricultural policies. Altering land and water rights through extending crop subsidies to certain crops or giving additional incentives for industries being set up in rural areas are only some of the examples of these shifts. Climate change has only aggravated these trends.

In Maharashtra for instance the post-2000s saw a lot of policy activity in the sector with the first water policy coming up in 2003 followed by the Maharashtra Management of Irrigation Systems by Farmers Act (GoM, 2005) and the Maharashtra Water Resources Regulatory Authority Act (GoM, 2005). These were introduced as part of the institutional reform proposed by the World Bank when the Maharashtra Water Sector Improvement Programme (World Bank, 2003) was launched in the state. The proposal also included setting up of river basin agencies in each of the five major river basins and bringing about an economic reform in the water sector. While the river basin as the planning unit was seen as important, there has been little of that in practice in the last 14 years since the water policy was introduced (Joy and Kulkarni 2010).

Reallocation of water

Almost every dam and reservoir originally built for irrigation is now used to also meet urban–industrial demands. Canal commands are increasingly unable to cater to their designed command areas, and this has a lot to do with reallocations of canal waters for other uses, as well as reallocations of lands within command areas for other uses. The main reasons for the poor performance of canal irrigation are as follows: (a) the conversion of canal commands into non-agricultural areas without the subsequent de-notification of irrigation commands being updated (updating avoided since the maintenance and repair grants are calculated on culturable command areas [CCA], so lowering the area means reduced grants for repair and maintenance), (b) diversion of water to non-irrigation uses (as high as 22 per cent in 2010–2011 for Maharashtra), (c) diversion of water from seasonal to perennial crops, which also happen to be water guzzlers (54 per cent of area under sugarcane is in canal commands of irrigation projects; Irrigation Status Reports 2005–2011), (d) upstream abstraction of water and (e) increasing use of canal water by abstraction through the ground rather than from flow (encouraged further by a policy initiative that has waived off the water charges on well irrigation in canal commands). All of these are giving rise to different forms of inequities for people within commands and also for those outside it.

Conversion of canal commands to non-agricultural areas: In several irrigation projects, command areas have been converted to nonagricultural areas. In some of the project commands, these have been converted to industrial zones as well. Unfortunately, there is no record of this as there are few ground-level studies that have been done by the Water Resources Department, along with the Revenue Department. Reducing the command areas on paper is also not a preferred option as that would reduce the budgetary allocation for repair and maintenance of canal systems which are still based on culturable command areas or the CCA.

Increasing use of irrigation systems for non-irrigation uses: As per the Special Investigation Team appointed to investigate into the several anomalies in the irrigation sector (GoM 2014) report, of the total water available about 22 per cent is used for non-irrigation uses and 20 per cent is evaporation, leaving only 62 per cent for agriculture. This was based on averaging of 10 years of data. This increasing use of irrigation quotas for non-irrigation uses such as domestic and industrial uses has been highlighted very well by a study done by Prayas (2011). They were able to show that over 1,800 million cubic metres of water were being diverted from agriculture to other non-irrigation uses. The assessment shows that more than 200,000 ha of land has been deprived of irrigation as a result of this diversion.[8]

These reallocations of water were done by a high-power committee set up by the government in 2003 which made arbitrary allocations to several industries and also to municipalities to meet the domestic water needs. Domestic water also increasingly includes water needs of different commercial enterprises which do not necessarily fit into the industries category.

Reallocations to commodity crops: Priorities in cropping patterns were also changing rapidly. The area under sugarcane, a water-thirsty crop, was fast increasing but more so it used more than 60 per cent of irrigation water.

The reallocation from seasonal to perennial crops has seen a definite rise in the canal system. In the last 15 years, the area under food grains has decreased and that under cotton and sugarcane has increased exponentially (Economic Survey of Maharashtra, 2015–16). Similarly, the area under cotton too was growing at an alarming rate. The area under food crops has gone down in absolute terms since the 1960s (GoM 2016–17). Thus, much of the growth in water appropriations in agriculture can be attributed to a shift from lower-value food crops to supposedly 'higher-value' crops like sugarcane or cotton. These cropping choices are determined by the market linkages and the profits that accrue to the political class in power. The water requirements of these crops are usually way higher than that of the food crops, however the benefits are not usually commensurate to the water used.

Upstream abstraction of water: Despite the progressive elements in the water policy of starting water rotations from the tail end of the canal to the head end and ensuring water supply and pricing on a volumetric basis, we have large instances of default. Some of this is partly due to the design of the irrigation systems themselves, as they do not allow for tail-to-head distribution, but a large part of it could be attributed to the political economy of water. The location of the haves at the head end of the system abstracts the water and thereby reduces the rights of those that are towards the tail ends, sometimes forcing them out of the systems.

Harnessing of surface water through the ground: This is the main focus of the present chapter and is elucidated in the examples that follow. The recommendations of all the earlier irrigation committees on the need to separating the two domains of surface and groundwater has clearly been ignored. Groundwater development in canal commands has increasingly led to the privatisation of the resource that has been generated through large public investments. Increasingly farmers in canal commands prefer to recharge their wells or to pump water in the farm ponds that have mushroomed through a policy of the government of Maharashtra, *Mageltyala tale*, which can be translated loosely as you ask and you get it – the farm pond. This allows them to escape the water charges that are levied on canal waters through gravity flow. With about Rs 500 billion spent so far and another Rs 750 billion requirement estimated to reach the ultimate irrigation potential, Maharashtra is very badly off in terms of the reach of irrigation. Of the total cultivable area of 225 lakh ha, only 13 per cent has been irrigated as per the ISR (2010–2011). Of this, only 8 per cent was through canals in 2010–2011 and the rest through groundwater. There is an increasing penetration of groundwater into canal commands in Maharashtra. Of the total irrigated area in canal commands, the area irrigated by wells increased from 26 per cent in 2000–2001 to 37 per cent in 2010–2011, while that irrigated by canals decreased from 73 per cent to 62 per cent in the same period. These shifts in uses of water are predicated upon the reallocations in land and water rights and, as in the examples discussed in this chapter, force the smaller and marginal farmers, most often the women, to exit as farmers and become wage labourers.

Water reallocations need to be understood in the current context of neoliberalism where the state plays an active role in facilitating and promoting such reallocations of water and water rights towards what are seen as more productive and more efficient uses. It does this through a series of water and land reforms, all firmly reflecting neoliberal ideas and premised on 'freeing' water and land by fixing entitlements for easing their transferability (see Ahlers 2010). The 2003 water policy in fact changed the water prioritisation and brought industry before agriculture. There was resistance to this, and finally, the government had to revoke the decision. The MMISFA act also allows for water rights to be transferred to the contract farmer in case of lands that are leased out for contract farming. Such policy and legal initiatives are creating a pathway for dispossession of the small and marginal farmers thereby compelling them to exit the system (Joy et al. 2014). There has been complete silence on the part of policy makers to respond to some of these ground realities, which is indicative of a strong backing of the state.

Moreover, policy makers have simply not been able to respond to this change as they have not been able to get out of the public infrastructure mindset, which has historically looked at irrigation provision through large infrastructure controlled by the state. The irrigation system is plagued by poor cost recovery, centralised management, political interference, weak bureaucracy and the lack of participation of users. However, what is insufficiently discussed is the burgeoning groundwater economy in the canal commands, which has changed the face of public sector irrigation. Increasing use of groundwater in canal commands has huge implications for equity, both in terms of direct access to water and decision making in its use and allocation, and also indirectly, in terms of changing land and labour relations.

To understand the impacts of the increasing use of groundwater in canal commands on water inequities, the case of the Kukadi major irrigation project is discussed in the following section.

The Kukadi major irrigation project

The Kukadi project is a multi-river, multi-reservoir and multi-canal project. Its 156,000 ha command area cuts across three districts of north (Ahmednagar) and west (Pune and Solapur) Maharashtra. It has five dams and a dense network of right- and left-bank canals, distributaries and finally the command areas at the minor level. In short, it has several levels at which water is distributed, and these levels cut across the political and physical boundaries of the region, making it a complex project to govern (Siddamal and Birajdar 2012).

The project was conceived in the pre-Independence period, and a part of the Kukadi canal was also constructed before Independence. The project however gained momentum during the severe drought of 1972 when the canal works started and also construction of the first dam at Yedgaon which is a collection weir. Political lobbying has also played a significant role in the way the project has evolved and changed its design over the years. For example, the second dam, Manikdoh, was to have six river sluices through which water from Manikdoh would be let out into the river and picked up at the Yedgaon weir. However, during construction, people from Junnar tehsil demanded that a small left-bank canal be constructed to irrigate their areas. This demand was fulfilled, and similar demands were then made by different groups of farmers backed by different political lobbies. This in a way laid the foundation of inequities in the Kukadi command.

Just as the construction designs kept changing, so did the cropping design. Kukadi dam did not have sugarcane as part of its cropping plan. The only crops for which irrigation was being planned were jowar, bajra (both hybrid as well as local varieties), groundnuts, vegetables and wheat – all were food crops. However, today we see a very different set of crops with different kinds of water requirements that cannot be catered to by the irrigation project. The scope of this is not to evaluate the Kukadi project with its mandated objectives but to demonstrate how groundwater use has pervaded the commands of the project and led to the concentration of water in the hands of very few people.

In 2012, SOPPECOM did a rapid overview of three irrigation projects in terms of the water entitlement programme that was introduced by the MWRRA in a phased manner after 2005. An entitlement is simply an authorisation to use a certain quantum of water. The MWRRA Act of 2005 mandates that the distribution of entitlements within the different categories of use after sectoral allocations are made by the Cabinet. The three main categories of use are domestic water supply, industrial water supply and irrigation. Bulk water entitlements for irrigation are to be issued by the river basin agency (RBA) to the water users' associations at the primary unit level or the minor level, distributary level and canal- or project-level associations.[9] Individual entitlements are issued for lift irrigation on reservoir or canals. Such entitlements are supposed to be administered, registered, measured and monitored by the respective RBAs in close coordination with relevant government agencies. Since the RBAs are not in existence, it is the Water Resource Department which functions as the RBA.

While MWRRA talks of entitlements in the context of the river basin, in practice these entitlements are limited to irrigation projects thereby ignoring groundwater. In Kukadi, the entitlements were worked out at the irrigation project level in around 2008–2009. It was one of the pilot projects of the MWRRA for its ambitious entitlement programme then. From 2008 onwards, the MWRRA itself monitored its entitlement programme for over a period of three years for Kukadi. The entire project is yet to be covered under this, but about 11 per cent, that is 8,029 ha of the first phase of its command, which is 67,000 ha, is covered under the pilot programme. For the years 2009–2010 and 2010–2011 actual irrigated area has been less than 15 per cent except for the rabi season of 2010–2011, when it was 33 per cent. Delivery of entitlements for Kukadi has been less than 50 per cent of the applicable entitlement, except for the hot weather (HW) season in 2010–2011 when it was about 66 per cent. If we look at the actual area irrigated and the total water delivered, it is about 1.5 times to three times the entitlement that has been worked out by the MWRRA for a normal year. For example, Kukadi water entitlement for normal rabi season is 821 m3/ha. Data shows that in 2009–2010, as well as in 2010–2011, water consumed has been more than 1,000 m3/ha and in fact in HW 2010–2011, it was about 3,000 m3/ha. What this indicates is that if the entitlement were to be equitably distributed across the entire command area, the per hectare share would be much lower than what the current practice is showing. Water is thus spread over a smaller area and is also concentrating in fewer hands. These inequities have grown over the years and have sharpened in the last two decades with the shifts in policies in water and agriculture.

From two locations in the Kukadi canal, this section shows the different practices employed by farmers to address poor outreach of canal irrigation. These examples show how the more prosperous and enterprising farmer is able to tide over the crisis at least temporarily, but the smaller farmers simply wait to a point that they are forced to exit the system. These examples are from a study by SOPPECOM (2012)[10] that looked at the command areas of several water users' associations (WUAs), including insights from two WUAs which are located on a single distributary of the Kukadi project in Ahmednagar district. They are located at middle and the tail end of the distributary and give an understanding of how water distribution is affected by the location and how people at different location scope with their locational advantages and disadvantages.

The Sahakari Pani Vapar Sanstha (Sahakari Co-operative WUA) is a tail-end WUA and currently does not receive any water from the Kukadi project despite being in its command. The WUA was formed 10 years ago, and although it does not receive any water from the system, it continues to function as a WUA and its 200 ha of command area is lush green despite the lack of canal water. The WUA has about 150 farmers owning about 1–1.5 ha of land on average. The main crops grown there are onion and pomegranate. The command area of this WUA has a large number of wells which benefit 90 per cent of the farmers for the rabi season. However, summer crops are grown by only the 20 per cent of the farmers who are able to pump water from the adjacent river.

The Dharmanath Pani Vapar Sanstha (Dharmanath Co-operative WUA) is a middle-reach WUA with a command area of 465 ha and a membership of 324 farmers. This middle-reach WUA is better off than the tail-end one. However, not more than 100 ha of it are covered by flow irrigation. Only about 15–20 of the 324 farmers benefit from flow irrigation. Like the tail-end WUA, this one had several wells in its command area: bore-wells that were below 300 ft and numerous dug wells. The landscape had a series of pipelines from adjacent streams to the farms. The entire command area was largely adorned with horticulture crops such as pomegranates and bananas apart from which there was also sugarcane and onion.

The area looked green despite the fact that the irrigation office showed a very low demand for water from canals. Most of the farmers in the command area of this WUA said that the frequency of water rotations has gone down in the last decade. In an entire year across the three seasons for more than a decade, they have not been able to get canal water for more than 10–15 days.

Discussion

Both the WUAs could thus not rely on the canal system to meet their irrigation requirements. Dharmanath WUA had a locational advantage and thus some hope that the system might improve and they could benefit from those improvements. The Sahakari WUA had no hope from the system as they were at the tail end of the distributary. This uncertainty of flow irrigation led to different kinds of changes within the command area. Those who could afford to set up their own systems prospered despite the uncertainty of canal flows, but the smaller farmers had to opt out of the canal system either by selling their lands or by leasing them out. Those with economic means laid pipelines from the streams to their farms, pumped water from the adjacent river or dug deeper for water. This provided them some assurance of water in a timely manner. Most of the wells in the command of this WUA are located in these streams, where water is freely lifted from canals or delivered through channels or pipelines made by the landowners.

A few prosperous farmers thus are able to harness canal water from a stream nearby. They do not demand water from the canal and thus do not pay for the canal water either. These farmers prefer to invest in wells in their farms and in the streams and in laying pipelines from the streams, which could be anywhere between 2 km and 5 km away from their fields. The streams are recharged by the canal flows thereby ensuring water security for the farmers. Some of those located at the head end of these WUAs are able to receive canal water, but the uncertainty forces them to soak their lands in order to ensure recharge of groundwater. The shifts in both the WUAs are clearly towards groundwater use. Several reasons were cited by the farmers; the most evident one expressed by them is the poor functioning of surface systems, which do not offer any water security to most farmers, thus forcing some farmers – mostly small farmers with no wells in the command areas – to move out of the system or those with wells and better resources to innovate and improvise, converting the public resource into a private one.

The irrigation officers from the senior to the lower levels cite political interference as a major reason for this mismanagement. Often local MLAs have pacified their constituencies by announcing expansion of command areas without paying due attention to the increase in numbers of farmers directly lifting from the canals, reservoir and the river. The junior engineer of the Kukadi project said that if one water rotation is completed in the entire command area of a minor canal, then several of the WUAs will not get even a single rotation in a season. Another official of the Kukadi project said that not more than 10 per cent of the area is covered in one rotation and not more than three rotations are usually possible in rabi and HW together.

Several Kolhapuri Type (KT) weirs have been constructed across the river which intercept the waters released from the dam at different points, thereby making the assured service to command areas very difficult. Non-irrigation uses for domestic water and industries have also seen a steep rise in the last two decades, thereby reducing the shares of irrigation. Water theft is not uncommon, and corruption is rampant. One of the junior engineers said that 50 per cent of the irrigation in the command area of the project is in fact serviced by river lifts and not by canal flows.

These are just examples of the innovations, alterations and adaptations made by farmers to survive in an uncertain irrigation system. The implications however are far more serious for those who are small and socio-economically disadvantaged, as they are often forced to exit the system.

Changing nature of the canal commands

The situation within the WUAs shows that the whole nature of flow irrigation is changing. These systems have been designed to operate independently of groundwater interactions. Today, groundwater has become an important element in the system, with potentially far-reaching implications. Only a small number of farmers earlier had wells. Now almost all farmers in the command area have wells. More and more farmers are relying on wells and looking at flow irrigation as a supplement and mainly as a means of recharging their wells. The uncertainty created by a lack of information, and inability to provide sufficient number of rotations due to over-extended commands, has exacerbated these trends in Kukadi.

Cropping patterns changed with better markets, infrastructural supports and subsidies offered to horticulture, floriculture and so on. Cereals were thus replaced by crops that promised better 'value' and required very high control over water. More efficient methods of irrigation such as drips and sprinklers have also meant that farmers prefer to receive as much of their water as groundwater recharge than direct flow application. This has led to profound changes in attitudes. Unlike flow application where head-reach farmers would release water after their fields were irrigated, groundwater preference has led to the tendency to accumulate water and head-reach demand has grown and lost the self-limiting character it had. This has made for sharpening inequalities between farmers at the head and tail reaches within the system at all levels.

Reallocations of water from subsistence or food crops to commercial crops and horticulture or from agriculture to non-irrigation uses alter the social relations as well, and this requires a deeper understanding. For example, there is a need to investigate further as to whether or not it has led to an increase in unpaid work of women and other marginalised groups. The unpaid work needs to be understood not only in terms of labour on farms or fetching depleting water for domestic uses, but also in terms of impacts on health due to the shift from food crops to commercial crops and the increased work burden of women in care and nurture.

In such a context, farmers/irrigators who can afford it often start modifying and adapting themselves to serve their own interests. The small and marginal farmers at the tail end of the canal commands often exit the system, are dispossessed of their land and water rights and end up providing farm labour, or worse still, migrant urban labour, and are dispossessed of their social and cultural roots as well (Srinivasan and Kulkarni 2014).

Those who can afford it exit the flow system but on their own terms and set up a parallel irrigation system by lifting through rivers or reservoirs or boring deep tube wells within commands, laying 2- to 3-km-long pipelines to bring canal water recharged in a nearby stream. Such examples abound the landscape of canal irrigation systems in India.

Conclusions

Surface irrigation with all its limitations at least partly can be said to be managed under the public sector where some formal rules of distribution and use can be laid down, however groundwater largely remains out of this purview. This move from surface irrigation to groundwater is a way to free up public water for unregulated private use. Access to groundwater depends on the ability to make investments in drilling wells and extract the water, leading to a concentration of entitlements and water rights in the hands of those who are economically and socially capable. Often these are marked by existing social hierarchies of caste, class, gender, tribe and other forms of social discrimination. Thus, the changing face of public sector irrigation warrants attention from policy makers and practitioners.

Changing discourse in water policy, prioritising private over public interests, is contributing substantially to these inequities. With the examples discussed it is evident that while the neoliberal framework has contributed to deepening the inequities, the state-led water development located in the political economy of water has in fact laid the foundation for the new forces to capitalise on the structural- and development-induced inequities.

The state plays a crucial role in backing and facilitating these processes, as it is responsible for modifying regulations or legal frames and institutions, favouring, privileging and even promoting such forms of capitalist accumulation as forms of economic growth, progress and development. The state, for instance, actively steps in to appropriate land and water resources for the benefit of cities and industry or within agriculture for commodity crops. For land, this is more directly visible than for water, but every time an allocation is made that transfers water from rural areas and farmers to urban areas and industry, dispossession is taking place. Indeed, these reallocations can even be understood as forms of water grabbing, supported by new powerful coalitions of interests (Wagle et al. 2012).

The water reallocations that we have seen from the previous discussion point to a broader process of capitalist transformation that characterise the current neoliberal context. Water enters the market and gets integrated into the capitalist system not by water grabbing by large corporations alone, but in diverse ways through reallocation of water use for commercial agriculture or directly for non-agricultural use. These reallocations are part of the well-thought-out strategies and policies of a state that enforces legislation to create individual water rights that are more amenable to trading and speculation. By doing so it renders those social sections of the society for whom water is more than just a 'resource' as helpless labourers with no dignity. It undermines their collective identities around water and also their long-term investments through their labour in farming and irrigating and fighting for their rights to land and water (Ahlers 2010).

There are many more sites of contestation against the current paradigm of water reforms and creative engagements with alternatives: struggles against dams (displacement, submergence); mass mobilisations for equitable distribution; innovative experiences in participatory irrigation management (PIM) that address concerns of equity, sustainability and democratisation; struggles of dalits and women for rights over land, water and forests for livelihoods and for the sustainable use of these resources; organic farming movements, right to food and basic services of health and water, sanitation and hygiene (WASH); struggles against privatisation and water reallocations; stakeholder processes and conflict resolution around water pollution; and innovative watershed development experiences. All of these reflect the widespread plea for a more explicit commitment to equity, justice and sustainable use in water entitlements and rights and for a more sustainable pathway for development in general. Efforts to address and understand water justice should creatively engage with these experiences to find pathways for restructuring the water sector along more equitable, socially just, environmentally sustainable and democratic lines.

Notes

1 This chapter was originally published as Seema Kulkarni, 'Canal Commands and Rising Inequity', *Sociological Bulletin*, Vol. 67, Issue 3, pp. 348–360. Copyright

2 The Phad system is a 400-year-old system where irrigation is provided on a contiguous block not necessarily owned by the people. This system was prevalent in northern Maharashtra and continues in some other districts too. See Datye and Patil (1987) for a detailed discussion on the Phad system.

3 Malgujari tanks (Malguzari tanks) were ponds made for water harvesting by the Malguzaars, who were zamindars or tenants in eastern Vidarbha, Maharashtra, two centuries ago. These tanks provided water for irrigation and also increased the availability of fish for local consumption.

4 Sir Visvesvaraya was an irrigation engineer par excellence. His contribution to canal irrigation systems in India has been phenomenal.

5 V. M. Dandekar was a renowned economist; V. R. Deuskar was a senior irrigation expert and Comrade Datta Deshmukh was a leader of the Lal Nishan Party (Red Flag Party). All three were part of the Drought Relief and Rehabilitation Committee instituted by the government, and this was popularly known as the 3-D Committee. It significantly influenced the response of social movements engaged in anti-drought movements.

6 The Chitale Commission has come up with a comprehensive set of recommendations, and it is rather difficult even to capture the most significant ones here because of space constraint. Hence readers are requested to refer to the report itself to get a full understanding of the recommendations.

7 For a detailed discussion on reforms in Maharashtra, see Joy and Kulkarni (2010).

8 Prayas study through the filing of Right to Information Act (RTI)'s collated data of diversion of water to non-irrigation uses in the period between 2003 and 2011. These decisions of diversion were taken by a high-powered committee set up in 2003 under the leadership of the deputy chief minister and also holding charge as water resources minister. These decisions completely flouted the law that had laid out a procedure for sectoral allocations of water.

9 MMISFA has made it mandatory for water users served by an irrigation project to form water users' associations (WUAs) at the smallest canal level, also known as the minor canal. All WUAs on different minors of an irrigation system federate at different levels such as the distributary and the branch canals to finally form a project-level association.

10 IWMI-Tata supported study – assessment of MWRRA entitlements and tariff, 2012.

References

Ahlers, R. (2010). Fixing and nixing: The politics of water privatisation. *Review of Radical Political Economics*, *42*, 213–230. doi:10.1177/0486613410368497.

Datye, K. R., Paranjape, S., Kulkarni, S., and Joy, K. J. (2004). The Krishna valley development: An approach to equitable and efficient use of water. In A. Vaidyanathan and H. M. Oudshoom (Eds.), *Managing water scarcity* (pp. 193–221). New Delhi: Manohar Publishers.

Datye, K. R., and Patil, R. K. (1987). Farmer managed irrigation systems: Indian experiences Centre for Applied Systems Analysis in Development, Mumbai.

Government of Maharashtra (GoM). *Irrigation status reports 2010–11*. Retrieved from https://wrd.maharashtra.gov.in/Site/1181/Irrigation-Status-Report

———. (1999). *Maharashtra water and irrigation commission report* (Vol. 1). Aurangabad: Maharashtra Water and Irrigation Commission.

———. (2005). *Maharashtra Water Regulatory Act 2005 (MWRRA)*. Retrieved from https://wrd.maharashtra.gov.in/Site/1169/Maharashtra-Water-Resources-Regulatory-Authority

———. (2005). *Maharashtra Management of Irrigation Systems by Farmers Act (MMISFA)*. Retrieved from https://wrd.maharashtra.gov.in/Site/Upload/Pdf/MMISF-ACT-2005.pdf?MenuID=1259

———. (2014). *Special investigation team's report on irrigation*. Mumbai: Water Resources Department.

———. (2016). *Economic survey of Maharashtra 2015–16*. Retrieved from www.maharashtra.gov.in/PDF/EcoSurvey_2015_16_Eng.pdf.

Joy, K. J., and Kulkarni, S. (2010). *Engaging with the changing water policy discourse in Maharashtra* (Globalisation Governance Grassroots Series No. 12). Pune: National Centre for Advocacy Studies (NCAS).

Joy, K. J., Kulkarni, S., Roth, D., and Zwarteveen, M. (2014). Repoliticising water governance: Exploring water re-allocations in terms of justice. *Local Environment: The International Journal of Justice and Sustainability*, *19*(9), 954–973. doi:10.1080/13549839.2013.870542.

Kulkarni, H., and Vijayshankar, P. S. (2014). Groundwater resources in India: An arena for diverse competition. *Local Environment: The International Journal of Justice and Sustainability*, *19*(9), 990–1011. doi:10.1080/13549839.2014.964192.

Lele, S. N., and Patil, R. K. (1994). *Farmer participation in irrigation management: A case study of Maharashtra*. New Delhi: Horizon India Books.

Purandare, P. (2012, July–August). Canal irrigation in Maharashtra: Present status. *Dams, Rivers and People*. Retrieved from https://sandrp.files.wordpress.com.

Shah, T. (2011). *Past present and the future of canal irrigation in India* (India Infrastructure Report). Retrieved from www.idfc.com/pdf/report/IIR-2011.pdf.

Siddamal, U. V., and Birajdar, C. A. (2012). Sustainable development of irrigation in Kukadi integrated project by transfer of surplus water from one reservoir to other. *WIT Transactions on Ecology and the Environment*, *168*. WIT Press. Retrieved from www.witpress.com.

SOPPECOM. (2012). *Maharashtra water resources regulatory authority: An assessment*. Paper presented at the IMWI-Tata Annual Partner's Meet, Ahmedabad.

Srinivasan, V., and Kulkarni, S. (2014). Examining the emerging role of groundwater in water inequity in India. *Water International*, *39*(2), 172–186. doi:10.1080/02508060.2014.890998.

Wagle, S., Warghade, S., and Sathe, M. (2012). Exploiting policy obscurity for legalizing water grabbing in the era of economic reform: The case of Maharashtra. *Water Alternatives*, *5*(2), 412–430.

World Bank. (2003). *Maharashtra Water Sector Improvement Programme 2003 (MWSIP)*. Retrieved from http://documents.worldbank.org/curated/en/139521468050066488/pdf/319970rev.pdf

Part IV

LAW, POLITICS AND RESISTANCE

10

RHINOCEROS IN KAZIRANGA NATIONAL PARK

Nature and politics in modern Assam

Arupjyoti Saikia

In September 2016, the government of Assam undertook a large-scale bureaucratic exercise to evict few hundred people from the neighbourhood of Kaziranga National Park. Those who were evicted were people living off agriculture, fishing or cattle breeding in the floodplains of the Brahmaputra River. The eviction was indeed a much-touted public affair. The state machinery used elephants, bulldozers, police and paramilitary forces to evict these few hundred families. The ensuring hostilities between the state machinery and alleged squatters led to the death of several people.[1] Much of this was the direct consequences of a 2015 Gauhati High Court judgment. The judgment was a result of several public interest lawsuits that were filed in the high court seeking eviction of alleged squatters from the park's newly defined boundaries. The judgment stated that Kaziranga National Park (hereafter KNP) should be made free from all possible human influence. The judgment had asked the government to make the park free from any illegal human occupation and also make room for expansion of the park's original boundary. The judgment had little doubt that 'the concept of national park in the Wild Life Act contemplates that there should be no human habitation'.[2] The tone of the judgment largely echoed the hegemonic public mood in Assam and received the appreciation of the protagonists of wildlife conservation. The judgment emphasised that villagers living on the fringe of KNP possibly connived in the poaching of wild animals:

> The individual claims for a handful of persons is in conflict with the public and national interest. There have been persistent and repeated reports of poaching of rhinoceros, elephants and other wild animals. It is irresistible inference that the habitants in KKP [sic] area would fall in suspect group and they would be well-acquainted with the areas and animal movements, therefore they

> would alone be in a position to do poaching successfully or abet poaching by others.[3]

The KNP administration however clarified that there were no squatters inside the KNP. Those who would be evicted were rather legal settlement holders or unauthorised settlers in the government lands. These government lands fall within the newly defined perimeter of the park. Indian laws stipulate that to redraw the perimeter of a park requires clearance of several environmental legal steps. While the government's views and actions regarding all these were largely obscure, the agrarian communities in the neighbourhood of the KNP were getting ready to spell out their rights in the environmental commons. Much of the ideological impetus for this emanated from a new era of political empowerment of agrarian rights within India's forestry programme after the landmark promulgation of India's Forest Rights Act of 2006.

As the Gauhati High Court pronounced its judgment, a general sense of uneasiness engulfed the neighbourhood of the park. This was soon followed by electioneering for the Assam Legislative Assembly in 2016. The main opposition Bharatiya Janata Party (BJP) left no stone unturned to make the poaching of rhinos in KNP into a highly charged election issue.[4] The BJP also promised to evict 'illegal squatters' from the territorial bounds of the KNP. These election promises paid off handsomely. After the election, in November 2016 the BJP-led Assam government declared its intention to fulfil its promises by evicting those encroacher families. In the run up to the eviction, residents refused to be shifted and energetically reasserted their claims to their lands and other natural endowments including fisheries and more.

The judicial indictment and political actions before and thereafter are a replication of bigger as well as more passionate public debates surrounding the questions of conservation and land rights in and around KNP. Increasingly a wider and more diverse set of institutions has joined these public debates to decide the future of the park and its surrounding landscape. Much of these debates are shaped by international wildlife conservation practices. The park is home to a wide range of fauna including the greater one-horned rhinoceros, perhaps the most endangered species of the Indian megafauna. In recent decades, there has been a major flow of international tourists to KNP who are increasingly captivated by the wild stories told about it. The park has also emerged as a major source of income for thousands of petty entrepreneurs as well as a handful of Assam's regional capitalists. In the poverty-stricken economic setting of Assam, the KNP offers a small hope for the poor. In recent years, the fact that the park has been recognised as a UNESCO World Heritage Site has elevated it to an important place in the body-politic of Assam.

What gets embedded into these public debates and political practices is the power of the cultural politics of Assam. The one-horned rhinoceros is available in only a few places in India, and the park for a long period of time was proud of its exclusive history of ownership, which had greatly shaped these cultural politics. What gets lost in this narrative of cultural politics is the complex ecological history of the Brahmaputra Valley, which plays an important role in making the ecological space of KNP a thriving one.

These questions are essentially and inescapably linked with two contentious subjects: the conservation challenges of the megafauna, including the flagship greater one-horned rhinoceros in particular,[5] and the lives of agrarian communities. Public debates, given shape largely within the Assamese public space, increasingly aim at alienating the agrarian communities from the landscape of the park. The governmental institutions also further reinforce this anti-agrarian public rhetoric. A series of legal and executive prescriptions try to sanitise KNP from all human influences. Over the decades, the park administration has also repeatedly articulated ideas of a shortage of space for increasing wildlife. This warranted overstepping the territorial limits of the park into agrarian space. Over the years, a set of legal and forestry prescriptions, which ignores both the landscape and ecological complexities of the region, has now set the tone for the governance of the park. These ideas are generally immune from the complex ecological, historical and political realities of Assam.

How does one understand the complexities involved in these public debates? This chapter tries to answer this. It begins by outlining the historical evolution of KNP, which needs to be prefaced by an understanding of the rhinoceros in an era of imperial hunting. The chapter then traces the administrative and political battles that went into the making of the KNP largely in a floodplain agrarian setting.

Towards an era of great hunting

Assam's pre-colonial landscape was not free from hunting. Royal hunting, illustrative of the grandeur of Mughal power, had a weak resonance but could hardly replicate it in its grandest form. A close reading of the contemporary Assamese records suggests the absence of large-scale violence against fauna. Neither did the local populace have extensive access to firearms. Rudimentary hunting traps, often made of bamboo, hardly gave the population enough advantage over the megafauna. Rather, some megafauna were trusted allies of Assam's medieval rulers. Elephants, like other Indian rulers, were the smartest animals whose importance in the making of the medieval polity of Assam is well-recorded. Political and economic reasons helped in the simultaneous growth of expertise on the care and husbandry of animals elephants or horses. Much of the expertise acquired sophisticated form and extreme labour was taken to put that knowledge on record. The best illustration was the *Hastividyarnarva*.[6] This mid-18th-century illustrated Assamese manuscript is perhaps one of the best examples to highlight the role played by elephants in the political landscape of medieval Assam. One can also cite the example of *Ghora Nidan* which incorporated details about local knowledge on understanding and care of horses.[7] Elephants and their tusks constituted an important portion of tribute to the Ahom royal treasury from their feudatory chiefs. Conflicts over the supply of elephants to the royal household had eventually led to three decades of political unrest in the late 18th century. This eventually led to the decline of the Ahom kingdom.[8]

In the floodplains of the Brahmaputra Valley, while everyone excelled in fishing, there was little to suggest that big hunting was practised on a large scale. Deer meat was exotic but nonetheless available for a wider section of the people. In the hills and foothills, innumerable ethnic communities keenly pursued a wide range of hunting practices. Their everyday life was tuned to the rhythm of small-scale hunting.[9] The primary motivation for hunting was for food. Also religious taboos restricted large-scale killing of animals. Absence of firearms also ensured that animal hunting remained confined to a limited few.

What changed since the mid-19th century was the arrival of big-game hunting in Assam. As the 19th century progressed, Assam, after the story of prospect of tea cultivation broke in faraway England, became part of a spectacular narrative of wilderness. Assam's fauna became a source of attraction for game and sport for many willing British who decided to travel to this far-off place. If a tea plantation could bring great speculative wealth, Assam's great fauna would help them to be in a familiar landscape of homeland. For many young British, the idea of the wild Assam was greatly attractive despite its remoteness from Calcutta. Unlike three presidencies in British India, Assam, despite her promise through the massive British investment in Assam's tea plantations, was a place far away from home. But her fauna and wilderness could still attract many. Each aspired to become a skilled hunter while in Assam. Assam's fauna came under tremendous pressure.[10]

In the decades to follow, while big animals were either ruthlessly killed or maimed, many escaped this cruelty. The most illustrative of them was the elephant. As the luckiest, the elephant became partner in the empire-building process. It helped earned substantial revenue. The number and variety of unlucky ones however was more widespread, though game hunting was not a very favourite activity in Assam mainly because of the soil condition of the region.[11] The attraction of Assam for sportsmen was best captured by R. S. S. Baden-Powell, who catalogued India's best hunting grounds:

> In Assam and Burma, as in many other parts pig is plentiful, but the ground impassable. On the Brahmaputra the pig are abundant, in fairly open country but as it consists for the most part of paddy fields, the ground is only passable in dry weather, and is then so hard, slippery, and fissured, that it is unrideable even to men like Colonel Pollok, accustomed to cotton soil.[12]

In the middle of the 19th century Major John Butler of the 55th Regiment of the Bengal Native Infantry found the sport in Assam an exciting pastime for the English sportsman. He wrote, 'from the vast extent of waste or jungle land everywhere met with it in Assam, there are, perhaps, few countries that can be compared with it for affording diversion, of all kinds, for the English sportsman'.[13] Butler catalogued a range of sports. He proudly claimed how in one day's sport it was no uncommon event for three or four sportsmen to 'shoot thirty buffaloes, twenty deer and dozen hogs, besides one or two tigers'.[14] Captain Pollock, a military engineer responsible for laying down the road networks in the Brahmaputra Valley in the 19th century, an anecdote claimed, shot dead one rhino or buffalo for every breakfast (Saikia, 2009). The Indian hinterland was richer than England in terms of the availability of game animals. Europeans were keen to experience the thrills of the chase and hunt. Encounters with big animals like the 'savage tiger' and the 'noble lion' were far more attractive and exciting than the routine business of spending small shots on birds. For James Forsyth, posted in India in 1857, 'the main attraction of India lay in the splendid field it offered for the highest and noblest order of sport, in the pursuit of the wild and savage denizens of its forests and jungles, its mountains and groves'. The range of the firearms of the colonial officers however may well have limited the impact of early British hunters on local fauna. Antelope shooting for instance could be only successful if the hunters got within 80–100 yards of the animals. The Assamese also across their class position participated in the hunting, as it was not merely confined to the higher echelons of the society, the poor too killed wild animals.

Some animals who made the floodplains their home encountered a great challenge. But there were also stories of survival. One of them was the wild buffalo. Integral to the Assamese rural landscape, and melodiously described in Assamese folklore, wild buffalo were increasingly seen in the official narratives of the British revenue officials as a big challenge to agriculture. Butler reported that in lower and central Assam large herds of hundreds of buffalo were frequently met with and the devastations committed on the paddy filed was incalculable. T. T. Cooper, a British sportsman in Assam, said of the wild buffalo, 'it was so numerous and so destructive as to be an absolute pest'.[15] But this did not necessarily act against the wild buffalo. What helped them to survive was the lucrative grazing economy. By the early 20th century buffalo grazing became a powerful component of Assam's rural economy. Buffalo grazing in the vast floodplains of the Brahmaputra had greatly transformed the rural economy by encouraging migration of Nepali grazers from the Terrai and allowing the flow of speculative capital into this economy. Some animals faced extreme hostility but required official protection. This was the case of the one-horned rhinoceros.

Almost everyone amongst the British officers took to hunting. Butler had no doubt that 'almost every military officer in civil employ in Assam, having constantly to roam about the country, becomes, if not from choice, at least in self-defence, a keen and skilful sportsman'.[16] Throughout the 19th century hunting in Assam and Burma became a major source of incentives for the British officials to take up a job in India. A. H. Meysey Thompson wrote in 1899:

> A young man in receipt of a comfortable salary can enjoy many things which are beyond the reach of any but the wealthy at home. . . . If he has a taste for sport, he can at very small expense spend his leave in shooting rhinoceros, buffalo, and bison.[17]

Rhinos and the imperial fantasy

The majestic character of the rhinoceros was known to the Europeans. Carolus Linnaeus (1707–1778), father of modern taxonomy, mentioned in his 12th edition of the *SystemaNaturae* (1766) that India was the habitat for the rhinoceros unicorn. The one-horned Indian rhinoceros definitely drew attention of the European naturalists prior to the 19th century. The rhinoceros drew the attention of the naturalists of Europe in earlier centuries. A number of naturalists between the 16th and 18th centuries either referred to India as habitat of the rhinoceros or paid special attention to its taxonomy. Carolus depended upon the texts of earlier scholars who reported on the Indian rhinoceros unicorns. It was surely in abundance in the first half of the last millennium in north and north-western India. A series of accounts between the 11th century and 16th century agree on the large-scale presence of the rhinoceros in India.[18] Accounts from the early 15th century indicate the presence of rhinoceros in the north Indian plains.[19] The Mughal accounts, especially those of the first emperor Babur, gave details of rhinoceros (*R. sandaicus*) including their hunting.[20]

Early British officials in Assam frequently reported the presence of rhinoceros. Their accounts contained details of these animals like their anatomical classification, ecological behaviours and also ecological distribution.[21] One such illustration was that of John McClelland, an assistant surgeon with the East India Company, who visited Assam in 1835 as part of the East India Company's appointed Tea Committee. Later he officiated as a superintendent of the Calcutta Botanical Garden from 1846 to 1847 and was also the editor of the *Calcutta Journal of Natural History* between 1841 and 1847. In 1841 McClelland prepared a detailed list of *mammalia* and birds collected in Assam and it was then that *Rhinoceros indicus* became listed as *mammalia* collected from Assam.[22]

But it was the stories of rhinoceros hunting in Assam and northern India was a major source of entertainment for British readers. Available British accounts of the hunting of the rhinoceros, largely unfamiliar to European eyes, are narratives of celebration of a megafauna which possessed enormous physical power. British tourists had taken immense pleasure in hunting and spotting the rhinoceros.[23] Specimens of the one-horned rhinoceros were sent from India to British natural history museums. In 1844, the Royal Institution of Cornwall received one skull of this animal which was sent by Assam's commissioner F. Jenkins along with other specimens.[24] Martin in his extensive accounts in *Eastern India* gave the earliest account of the rhinoceros. He reported that rhinoceros was found wherever 'there are forests and extensive thickets of reeds'. In northern Bengal, he reported, many earned their livelihood by hunting this animal. He thought that the animal was 'quite harmless, and neither injures the persons nor crops of the inhabitants'.[25] The *Oriental Sporting Magazine* regularly published several pieces on rhino hunting in Assam. Such hunting expeditions often met with unexpected challenges from other big animals and the complex ecosystem. The hunters normally roamed around grasslands and river islands, crossed flooded rivers and swampy lowlands and needless to say they travelled mile after mile to chase their prey. Despite large numbers of rhinoceros being killed, even by the 1880s, rhinoceros were still 'plentiful in some out-of-the-way districts' wrote George M. Baker, a tea-planter, in 1884.[26]

However, even with the rapid spread of hunting in Assam, the rhinoceros still had some defence. How was it possible? Butler was also of the view that 'rhinoceroses are very numerous in many parts of Assam, and are to be found in very high grass jungle, near inaccessible miry swamps, which preclude pursuit, and having thick skins'.[27] Not everyone could hunt the animal. Butler agreed that 'they are not easily shot'.[28] The Assamese supposedly had a fear of them. 'Elephants dread the charge of a rhinoceros as much as that of tiger, and the grunting noise of the former animal not unfrequently scares even a well-trained elephant from the field'. The elephant had every reason to be worried of the rhinoceros. 'If the rhinoceros succeeds in overtaking the elephant's sides or legs, and with the horn on the nose not unfrequently inflicts fearful wounds'.[29]

One of the most celebrated and addicted rhino hunters was the Maharaja of Cooch Behar. Between 1871 and 1907, the Maharaja had claimed he shot dead 207 rhinoceros.[30] In 1899 the Maharaja, along with several members of the British aristocracy and his own family, shot dead five rhinos on a half-day on March 7.[31] These were instances of those hunting expeditions whose information found space in the world of newspapers. There must have been many more. The animal, due to its majestic demeanour and the uniqueness of the horn, began to be displayed at royal ceremonies. The Colonial and India Exhibition of 1886 displayed 'shields of rhinoceros skin, with gold and steel bosses, and steel shields curiously damascened'.[32] One over-enthusiastic Indian prince sought permission to parade a rhino in the Delhi Durbar of 1903.[33] But by this time there was significant mercantile interest in rhino horns. J. Errol Gray, a tea planter in Assam who toured the hilly regions into further east of the Brahmaputra Valley, came across a thriving trade in the rhino horn between Marwari merchants and the Singpho tribes.[34]

There were more painful accounts of rhino hunting. M'cosh, the British officer mentioned how

> the mode of taking them [young calves] is first to shoot the mother, and then the calf is easily secured. Frequently, the mother, in her dying agonies, lays hold of her young one with her teeth, and lacerates it so severely that it dies of its wounds.

If the increasing numbers of hunters and sportsmen were factors in the decline of the rhino population, the government was equally responsible for the annihilation of the animal. Baldwin mentioned a government 'reward of twenty rupees to anyone shooting a rhinoceros'.[35]

By the 19th century one-horned rhinoceros (*Genda*, *Gonda*, *Gor*) presence in India was well documented but their retreat from several parts of the country also came to be noticed in a wide number of writings. While we do not know the total numbers of rhinoceros killed in the 19th century, one such report claimed in 1868 that hunters killed 'no less than 200 rhinos in the Goma Dooar'.[36] Some of these reports outlined a sense of the animal being on the verge of extinction; the rhinoceros 'will soon be exterminated'. Reporting their presence across 'Terai, and country between the Himalayas and the Ganges; from Rohilakund (Jerdon) in the west to Assam', William Lutley Scalater, the British zoologist and superintendent of the Indian museum in Calcutta reported that 'it was formerly plentiful in the Purneah district, but now seems to be almost confined to the Doars to the east of the Teesta River'.[37] Pollock during his many years of encounters with the rhinoceros in the foothills of the Himalayas and in the Brahmaputra Valley wrote how 'they are very plentiful in certain localities . . . the Terai, at the foot of the Bhootan range, and are also the swamps along the base of the Cossyah and Garrow Hills.' In 'Assam they inhabit the churs or islands in the bed of the Brahmapootra river.'[38]

The practice of hunting of rhinos was rampant in Assam and Bengal. Curzon himself shot numbers of rhinos in the early years in Assam and Nepal. In April 1901, Curzon wrote to his wife, then in England, about his total tally of trophies. 'Altogether twelve tigers were shot besides the rhinoceros and numbers of deer, boar, partridges and florican'.[39] In another time, his biographer informed, he chased and shot a rhino in an almost impenetrable jungle of pampas grass,

> I saw the great brute dimly standing in a sort of tunnel that he had forced for himself through the bottom of the grass. He turned and fled. I fired a shot that caught him in the neck and sent him over like a rabbit. Then you never saw such a commotion. He kicked and

> plunged, and we had to pour at least a dozen shots into him before he was finished off.[40]

Viceroy Lord Curzon again visited Assam in 1905, after the redrawing of the maps of Assam and Eastern Bengal, and he spent time hunting game. His 'inroad into one special rhino reserve (Kaziranga) resulted in a bag of only one cow rhino, where twenty years ago, in and around this same spot, fifty could be seen'.[41] Rules regarding big game were already coming in force but those rules were not meant for the Indian ruling elites.

Meanwhile, there was also some change in the attitude of the higher echelons of the British government in India. The most representative example was Lord Curzon himself. A skilled hunter, Curzon himself began to appreciate the need for the preservation of fauna. But he was careful not to overemphasise what he wanted to achieve. In 1901 Curzon was visited by the Burma Game Preserve Association, mostly consisting of Britishers. Curzon was on a visit to Burma. The association had drawn attention to the annihilation of the fauna. Curzon in return decided to speak at length on this subject through a public talk.[42]

Though big game rules were already in place, the increasing scarcity of big game including the rhinoceros still caused alarm. A popular British magazine on games and hunting, *Forest and Fauna*, wrote in 1909 that the danger to big game in Assam was still not over. 'The man that is doing all the damage and driving rhino and tiger from their old breeding places is the Nepali graziers [sic]'.[43] The journal bluntly and strongly accused the government of allowing the graziers to settle in return for payment of tax. The government 'allow these men [Nepali graziers] to bring in their large herds of buffalo at their own sweet will over every well-known breeding spot in Assam – grazing, and hacking down the best covers in the province'.

Sportsmen were still happy that there were places to hunt, but there were only a few places where the rhino could be hunted:

> On the south bank of the Brahmaputra there are still a few breeding places left, but these are inaccessible till about April, after the jungle has been burnt off. The government reserve forests in these parts can hardly boast of a rhino inside the areas.

The chance of big game, including the rhinoceros, was still a matter of temptation for the prospective British tea planters in Assam even in the last quarter of the 19th century.[44] By the early 20th century on the south bank of the Brahmaputra the homes and feeding grounds of the rhino essentially remained limited to lower ranges at the foot of the 'Mekir and Garo Hills, in the low lying swamps and dollonies, covered with dense ekra and kagri, almost as thick as bamboos, growing [to] a height of from 15 to 20 feet, through which it is impossible to drive a hathi at the pace of a rhino'.

Despite the imposition of restrictions on rhino hunting, other areas, particularly the rhino-inhabited places on the north bank of the Brahmaputra, remained exposed to rhino hunting and the hunting continued. In 1905, as the prince and princess of Wales were preparing their visit to India, the *Sevenoaks Chronicle and Kentish Advertiser* excitedly reported the grand preparations for the tour and did not forget to mention that 'it is not unlikely that the Prince may also be shown some rhinoceros shooting in Assam'.[45] Lord Minto, while visiting Assam in February 1909, shot one rhinoceros. The accompanying photographer, who happened to be the Viceroy's military secretary, fractured his leg as the rhinoceros charged the elephant on which they were riding.[46] George V, soon after his coronation, arrived in India in 1911 and took a hunting expedition to Nepal where he shot eight rhinoceros.[47]

By the early years of the 20th century, the rhinoceros came to be found only in the Brahmaputra Valley and a little to the west of this. In 1903 the *Sevenoaks Chronicle and Kentish Advertiser*, while reporting the receipt of an 'unusually fine specimen of the Indian rhinoceros' in the Natural History Museum, which was endevouring to get a specimen, also decided to mention that the animal 'is now to be found only in a small part of the Assam jungle, in the Maharajah's territory (Cooch Behar)'.[48]

Three decades later another British newspaper boldly asserted that 'the Indian rhinoceros has been driven from various parts of India and is now found only in the jungles of the Assam plain and Nepal'. Even then the animal was still a rare one in Nepal and was considered as royal game and 'its life is preserved by the severest laws'. Not only that, 'the Natives who dare to shoot the Royal rhino of Nepal are liable to heavy fines, and in some cases the penalty is death'.[49]

Writing in 1921, E. P. Stebbing explained this extraordinary confinement of the rhinoceros to a limited geography: 'The great opening out of the country, clearance of forests for tea gardens, cutting them up by roads and railways have restricted the area of untouched primeval forest which forms the natural habitat of this animal'.[50] In fact, in 1902, Francis Henry Skrine of the Indian Civil Service, while talking about the growth of the tea plantation in Dooars, showered heaps of praise on the tea plantation for forcing away the rhinoceros:

> If the man deserves praise who makes two blades of grass flourish where one grew before, of what reward is he not worthy who has converted a haunt of the rhinoceros and wild elephant into the chosen home of a great English industry?[51]

In 1932, the South Kensington Museum displayed an exhibition of 'game animals' and commentators spoke out regarding how the exhibition 'must be a revelation to people who do not realise how rapidly the creatures of the wild are being extinguished'. C. W. Hobley, secretary of the Society for the Preservation of the Fauna of the Empire exemplified the case of the Indian one-horned rhinoceros, which met with a similar fate.[52] 'Take the case of the one-horned Indian rhinoceros, once so common in India that herds were said to be used in battle in medieval times. The merest remnant now survives in the forests of Assam, and in Nepal'. Hobley sounded the death alarm and took the case of extinction to the international audience, he was hopeful that 'inviolable sanctuaries' would be formed 'in Africa, India and elsewhere, so that the vanishing wild life of the Empire may find peace and multiply'. The wild animals should be 'objects of travellers, and not merely targets for marksmen – victims of men's avarice and women's vanity, and at the same time, without interfering with legitimate economic development'.

Saving the rhino

By 1900 rhinos were found only in Nepal, Bengal and the floodplains of the Brahmaputra.[53] This happened due to a range of issues including disappearance of habitat and hunting.[54] The animal was not hunted and killed merely as a trophy. By the late 19th century Kaziranga was a planters' and other sportsmen's heaven. E. P. Gee, who had first-hand experience of the game reserve in its early days, wrote:

> in 1886 a certain sportsman went out on elephant in the area, which is now Kaziranga to shoot rhino. He encountered one and fired about a dozen shots at it from a close range. The wounded rhino made off, and as it was too late in the evening the hunter returned to his camp. Next day he followed up the bloody trail of the badly wounded rhino and came across it while it was actually engaged in fighting and keeping off two tigers. One tiger, the account says, had his neck fearfully covered with blood. The sportsman fired at both the other tigers, which escaped, and then finished off the unfortunate rhino.
>
> (Gee, 1952)

For hunters and poachers it was easier to follow the animal, as it left behind traces, and its horn was in great demand. The animal's horn was 'reputed throughout the East to possess aphrodisiac properties'.[55] Rumors about medicinal properties of the rhino horn had intensified its trade. E. P. Gee blamed this on the 'fanciful belief in the wonderful properties of its horn'.[56] The Javanese rhinoceros disappeared from Lower Burma by the late 19th century and this accentuated the pressure on the greater one-horned rhinoceros found in the floodplains of the Brahmaputra. The lucrative trade in rhino horn attracted the entry of merchant capital. Calcutta-based businessmen formed an organisation to make arrangements for the regular supply of rhino horn and elephant tusks to Calcutta.[57] Milroy reported that local Bodos living in the neighbourhood of the Manas game sanctuaries had taken to poaching on a large scale.

As pressure on the rhinoceros increased in the last years of the 19th century, it sent alarm bells ringing.[58] One of the earliest pieces of official communication was authored by J. C. Arbuthnott, a senior official posted in Assam. Arbuthnott's official letter of 1902 to Assam's chief commissioner portrayed a very grim picture of the fate of the one-horned rhinoceros. He wrote, 'the animal which was formerly common in Assam, has been exterminated except in a very narrow tract of country between the Brahmaputra and Mikir Hills in Nowgong and Golaghat where a few individuals still exist'.[59] Arbuthnott made it abundantly clear that 'reckless and indiscriminate destruction of all game' was done by 'large shooting parties from Bengal'. What was required was an official order to stop 'destruction of rhinoceros in Assam by shooting or by pitfalls' unless one did not wish to witness the 'complete extinction of a comparatively harmless and most interesting creature'. While he did not make it clear who constituted these shooting parties, the tone of his letters tells us that they were the natives, not the Europeans who hunted this animal for trophy. Henry Cotton, Assam's most illustrious chief commissioner, did not disagree with his junior officer but noted his reservations against keeping vast tracts of land which might be taken up for expansion of agriculture. 'It would not be desirable that the formation of the reserve should prejudice the development of cultivation, but there must be extensive areas suitable as habitats for the rhinoceros which are quite unculturable', Cotton's secretary conveyed to Arbuthnott a month later.[60] Cotton was famously known for his zeal to bring thousands of miles of vast wasteland in the Brahmaputra Valley under cultivation.[61]

The hunting of the rhinoceros was already prohibited in Bengal. This helped Assam officials to fall back on the protocols to be followed if something similar had to be enacted in the province. The lead was taken by Assam's civil officials much before the Forest Department had anything to say on this matter. A few years of official correspondences finally paved the way for enacting the legal and bureaucratic mechanisms to ensure a space for the rhinoceros. Arbuthnott, with the help of several junior British officials, including Major P. R. T. Gurdon, who would shortly make a name for his contribution to the Assamese language and cultural history, identified three vast spaces with an aim to establish rhino asylums in the floodplains of the Brahmaputra. The parameter for selecting these areas was clear: they should not stand in the way of the expansion of agriculture in Assam. Gurdon thus selected the future Manas National Park: 'The whole area does not contain a single village, for people will not live there for fear of the Bhutias. The Kacharies have no rights in the proposed reserve, but they go up sometimes to fish and also to shoot'.[62]

Arbuthnott's next choice was the vast stretch of grasslands in the floodplains along the Brahmaputra. 'In Nowgong I recommend the formation of a reserve to the west of Laokhowa and north of Juria between the Leterijan and the Brahmaputra river and in the Golaghat subdivision of Sibsagar in the vicinity of Kaziranga'.[63] What did both these places look like? Though not completely free from human movement, Arbuthnott thought that he could very well demarcate 'at both places a suitable area of uncultivated and uncultivable waste, destitute of inhabitants . . . without prejudice to cultivation' as asylum for the rhinoceros. As these official correspondences were heading towards the intended goal, J. Bampfylde Fuller, Assam's chief commissioner also lamented that the Forest Department 'have not done apparently much towards affording an asylum to the wild fauna of the country'.[64] Fuller also thought that whatever sport rules existed in Assam were a dead letter. The Assam government moved fast to declare those areas as reserved forests with an additional aim to make them rhino asylums. However, it made it amply clear that declaration of such asylums should limit the 'amount of game to be shot', prohibit 'the shooting of immature animals or females', limit the 'number of elephants to be taken into the forest' and that a licenced shooter would be accompanied by 'a forest official'.

It is most likely that the local Assamese elite did not welcome this idea of preserving forest lands as exclusive sites for the rhinoceros. J. Donald, Deputy Commissioner of Sibsagar, spoke to a few such individuals. They were all 'acquainted with the tract, and have been shooting therein'. They all opined to Donald that 'the tract should certainly be reserved in order to prevent the extermination of the rhinoceros'.[65] There were scattered families 'north of Diffloo river and south of Mariahati Mirigaon' but Donald was of the opinion that those families could be easily relocated.[66]

The rhino asylum was born amidst a general cry for lost species such as lions and bison and also the imperial rulers' claim for privileged hunting practices.[67] The makers of Kaziranga, being essentially both foresters and revenue officials, paid little attention to the complex landscape and agrarian practices.[68] The fact that its landscape is formed by the complex of sprawling floodplain grasslands, numerous water bodies and woodlands that provide an ideal mix of habitats for a variety of flora and fauna was admitted by the Assam officials. The Assam government made the quickest choice by selecting a long strip of land – mostly covered with savannah grasses and swamps – that could be easily accessed by the Europeans.

These strips of land across both the banks of the Brahmaputra were part of the dynamic floodplain system. The continuous making and unmaking process was an essential feature of this landscape. When the 17th-century Mughal traveller Shihabuddin Talish travelled through the Brahmaputra, accompanying the Mughal commander Mirjhumla, he saw human settlements and a dense agrarian landscape in the larger areas where the park is presently located.[69] Subsequent to Shihabuddin's departure, the Brahmaputra changed its course and large parts of these areas were transformed into swampy, low-lying lands, making it unfit for agriculture. In 1904 when J. Donald, deputy commissioner of Sibsagar, made local enquiries he had no doubt that the vast swampy areas of the Kaziranga could have been easily converted into cultivable land had there been some attention.

> The land might be made fit for cultivation if the old embankment, constructed in the time of the Assam Rajas, were repaired, but this would prove an expensive undertaking, and in view of the fact that the country is infested by wild animals the question of repairing the embankment need hardly be considered.[70]

Maps drawn by British officials during the later decades of the 19th century had repeatedly noted Kaziranga. Most beels (small lakes) and other wild areas in and around Kaziranga were identified with names. Such names were probably given by the erstwhile villagers who frequented these places when they were either densely or thinly cultivated. These floodplains of the Brahmaputra were witness to many volatile geo-morphological transformations. The retreat of humans took place here as nature was an unwilling host leaving the wild animals to swim, graze and roam around the wilderness.

It took some time to remove doubts over rights of the villagers and resolve issues related to the restriction of shooting privileges of the Europeans in India. Bampfylde Fuller was particularly unwilling to make any announcement that was hostile to the Assamese *raiyats*. Categorical voices came from the pen of the revenue officials. One of their central concerns was to make a revenue surplus province. They could not afford to be seen as hostile to the idea of the expansion of cultivation, a field where the province had been doing badly. Assam's tea plantation had paid off handsomely but the province could still accommodate people and newer species of crops. Most officials posted in Assam were also aware of the floodplains' wilderness. These officials, many of whom had experience working in the Bengal delta, had come to accept these floodplains as predominantly characterised by transitory human settlement. This made them careful not to dismiss the absence of the permanent and thickly inhabited village as the only proof of sign of human control of these areas.

It took a few years more to finally promulgate Kaziranga as a game reserve in 1908.[71] This meant there were restrictions on the shooting of animals, especially the rhino. The game reserve probably helped slow the reckless killing of the animal but hunting and poaching did not stop. In February 1909, Colonel Victor Brooke, military secretary to Lord Minto, India's viceroy, shot at rhinoceros when Lord Minto paid a visit to Assam.[72] There was increasing international demand for the protection of wild animals and Assam, as the only place for the near-extinct rhinoceros, now drew international attention. P. Chalmers Mitchell, president of the zoological section of the British Association stated that by the early 20th century the rhinoceros had been nearly exterminated from northern India and Assam.[73]

Until well into the mid-1930s poachers' camps were found at 'every bheel' and 'about forty carcasses of rhino with the horns removed' were noted by a forest officer who was deputed into the Kaziranga for survey.[74] Meanwhile, there was a sudden rise in rhino poaching in Assam in the 1930s. There was an increased international demand for rhino horns, often fetching the price of rupees 15 a tola.[75] Poaching had intensified in Assam. The government had sought the 'intervention of the troops of the Assam Rifles'.[76] Later conservationists believed that this armed intervention 'resulted in greatly slowing down rhino poaching and in more or less maintaining the position, at least within the reserves'.[77] At a moment when the rhinoceros were hardly at the forefront of governance, protection by rifles was a welcome move.

After about two decades, in 1926 the game reserve came to be known as a 'game sanctuary'. This meant that it still did not shy away from the social imagination of the trophy hunters who still considered this as a place for game. It was only towards the late 1940s that it came to be called a 'wildlife sanctuary' implying the inherent idea of all living creatures and their intended conservation.[78] Things have changed fast since then. Tourists slowly arrived but forest officials were never enthusiasts. Gee recalled an encounter with a British forest official who categorically told him that 'no one can enter the place. It is all swamps and leeches and even elephants cannot go'. Probably the British forest official was not wrong.

But the idea of game reserve did not win the sympathy of the neighbouring population, who were worried about crops being destroyed by the increasing numbers of animals. Neither humans nor animals had learnt the art of appreciation of each other's camaraderie. Only in 1924 villagers from the neighbourhood of the sanctuary wrote in a strongly worded petition to the Forest Department how thousands of villagers and their crops were being badly affected by the increase in wild animals. They wrote how animals from this game reserve had destroyed their crops and killed people. The villagers further appealed to the Forest Department to allow them to cultivate lands within the game reserve so that they could further push the boundaries of the wild animals away.[79] The villagers' uneasiness about the increasing population of wild animals however could not convince the Forest Department to think similarly.

The game sanctuary was opened to visitors in 1938. E. P. Gee was one of the first to come. Riding on an elephant, he saw rhinos who 'appeared to be most improbable-looking and prehistoric-like with their quaint features and thick armour plating'. It took another few decades for Kaziranga to gain attention. A few forest officials undertook significant steps to ensure that Kaziranga remained free from poachers. While many had probably worked tirelessly, the works of Mohi Chandra Miri, an assistant conservator of forests, and A. J. W. Milroy, conservator of forests in Assam in the

1930s, had crucial significance. Gee recognised Miri for clearing Kaziranga of pit-digging poachers.[80] The rhino also received the blessings of the Assam Legislative Council in the mid-1930s when the rhino horn was declared as forest produce.[81] This gave forest officials the right to seize rhino horns and provided a little respite to the rhino. However, things were still gloomy. Trade in the rhino horn was in full swing. Yet the physical well-being of the megafauna was 'largely a matter of individual whim'. Milroy had little doubt that 'encouraging results obtained by one Divisional Forest Officer are only too often dissipated during the regime of successor, who is indifferent to this side of his multifarious duties'.[82] Another contemporary observer squarely put the blame on the Indian junior officials: 'poaching goes on extensively under the eyes of Indian subordinates who are supposed to look after them'.[83]

Meanwhile there were some dramatic transformations in the floodplains. This brought wild animals and man more into direct confrontation. By the early 20th century, these floodplains fast emerged as a site for buffalo grazing.[84] This commercial venture was supported by the local elite. The buffalo herders' obvious choice was the riverine localities or the river islands of the Brahmaputra. These areas are famous for the growth of 'enormous areas of grass and reeds' and were most preferred by the megafauna. The large-scale growth of buffalo rearing came in direct confrontation with the habitat of the rhinos and other megafauna. Buffalo rearing had lost its economic and speculative power by the mid-20th century but not before producing a temporary setback to the idea of promised space for the rhinoceros. Meanwhile, habitats for megafauna decreased further. This additional pressure came simultaneously but differently. The central protagonist here was the jute crop. Jute was the most thriving agricultural crop in late-19th- and early-20th-century India. Beginning in 1860s Bengal, it had developed into a powerful mechanised industry which was ever ready to consume millions of bales of jute at regular intervals.[85] To fulfill the needs of this Bengal industry, by the early 20th century the crop began to be cultivated large-scale in the floodplains of the Brahmaputra.[86] The floodplains of the Brahmaputra were a great site for commercial agriculture by the mid-20th century. This had multiple consequences for the megafauna of which habitat shrinkage was the most crucial.

If the megafauna and peasant agriculture came into direct confrontation by the mid-20th century, there was more to it. The massive land resettlement programme that the British colonial government undertook since the 1890s with an aim to enhance land revenue received a jolt after the 1950s. After decades of uncertainties, the land revenue settlement process began only towards the last decades of the 19th century. It remained an unfinished task of the imperial government even in the early decades of the 20th century. Several factors assisted in making land settlement an incomplete process. The intimidating floodplain landscape, where one has to make a careful choice in selecting lands for habitation and agriculture, and peasants' practices of mobility and freedom to reclaim lands continuously acted against the much-desired wish of the colonial government. Further, the colonial land settlement programme was embedded in the flexible features of the floodplains. This ensured that the land settlement could never be completed. Trouble occurred when this unfinished state initiative encountered the complex and combined challenges emanating from Assam's tea plantation and Forest Department. The later decades of the 19th century were a time for the consolidation of the British tea plantations in Assam.[87] Similarly, the rapidity and hurried manner in which the Assam government handed over lands to the Forest Department did not match with the increasing needs for agrarian expansion necessitated by the rapid population growth since the 20th century.[88] Estimates from the early 20th century indicate that both the Forest Department and the tea planters combined owned half of the valley's land. All these factors significantly restricted the mobility of the agrarian communities. The late 19th and early 20th centuries were also a time for the retreat of the animals in the Brahmaputra Valley. Multiple agencies, including the tea plantations, the mono-plantations protectively promoted by the Forest Department and rapid expansion of the jute cultivations massively fragmented the floodplain landscapes. With the fundamental redrawing of the landscape architecture, animals' habitats shrunk, their routes of migration were disrupted and they were confined to limited ranges. Traditional elephant 'corridors' were broken up, habitats of other large mammals were pushed back and smaller varieties were made to disappear. Elephant habitats (elephants being the most important source of power and revenue both for the pre-colonial and imperial government) were pushed to the eastern and southern frontiers of the valley, which were yet to be touched by the advent of modern capital. When the floodplains were reclaimed at an extraordinary speed for jute cultivation in the first half of the 20th century, animals had to retreat from the riverine areas and sandy river islands.[89]

Science, rhinos and becoming a cultural heritage

Assam's cultural investment in the Kaziranga began only in the latter decades of the 20th century, but a similar project with the rhino preceded long before. Animals often occupied a significant place in Assam's political and cultural landscape. The finest illustration of this was the role played by the elephants in Assam's cultural and political history until at least the 19th century. The Ahom rulers jealously guarded their rights over the elephants from that of the Mughals. In the wake of aggressive British assertion to claim exclusive ownership of the elephants, the zamindars of Goalpara tried to retain their privileges in the last decades of the 19th century but they could hardly succeed.[90] Elephants acquired new political and cultural meanings through these histories of contestation and competing claims, but that hardly matched with the tragic narratives associated with the life and times of the rhinoceros.

The rhinoceros offered something more in the second half of the 20th century. Two distinct developments began to shape the future of the rhinoceros: the political-cultural commodification of the rhinoceros as well as the slow arrival of the discourse of conservation science. As early as 1948 the Assam government, jealously guarding the linguistic and cultural uniqueness of the Assamese identity, declared the rhino as the state insignia. The Assam State Transport Corporation, the government-run public transport facility, incorporated the rhino as its emblem; its red and black buses proudly displayed the rhino insignia. The Assam government also gifted, donated or presented rhinoceros to zoos, foreign governments or natural museums; some of those were christened with names drawn from Assam's cultural heritage. A rhino baptised *Lachit Barphukan*, after the medieval Assamese military commander, was gifted to Bombay Zoo in 1952.[91] The National Zoological Park in Washington received its first one-horned rhinoceros in 1963. Several foreign nations also bought rhinos at a cost. In 1967 a rhino was priced Rs 1 lakh for foreign buyers.[92] Most rhinoceros had to overcome a strenuous long-distance journey and adjust to a new environ. Their story of survival became enigmatic and reaffirmed their unique position in the cultural history of fauna. Cultural reproduction and global circulation created an unusual profile of the rhinoceros in the second half of the 20th century. Since then many joined in the remaking of the cultural image of the animal; the state, naturalists, peasants, small entrepreneurs and lawmakers were all part of this effort.

Before the territorial limits of the Kaziranga become part of Assam's natural and cultural heritage, a number of naturalists began to reimagine the future of the rhinoceros in the late 1940s. At the request of the Assam government three naturalists made their maiden visit to the Kaziranga in 1949.[93] Two of these naturalists were stalwarts in their fields. They were S. Dillon Ripley, a young naturalist then teaching zoology at Yale University, and the well-known Indian ornithologist Salim Ali. After their short trip to KNP, they suggested to the government of Assam that the Kaziranga could be 'turned into a show place and field laboratory'.[94] This team of naturalists agreed that Assam's game sanctuaries were of great importance as these were vast stretches of space which was 'the very last refugees of the Great One-Horned Rhinoceros'. They however agreed that poaching still existed and much needed to be done to ensure the survival of the rhino. E. P. Gee rightly claimed that their report 'exposed the optimistic complacency which had hitherto prevailed concerning the rhino population'.[95] Ripley and Ali's report caught the imagination of the influential *Times of India*,[96] which published a long editorial on this report and what was to be learnt from it. There was little doubt that wild animals in this game sanctuary were still a source of threat; rhinos occasionally straying out of the sanctuary and killing people continued to make international news.[97] In the 1940s a number of rhinos died of anthrax, an infection caused by bacteria. These deaths, while they did not cause any immediate alarm, would become a source of major concern in the near future. But there were competing claims about the number of rhinoceros after four decades of protection. Compared to the claims of a sizeable numbers of rhinoceros, Gee, much to the dismay of the Assam government, reported in 1950 that there were an estimated 150 rhinoceros in Kaziranga.

The Bombay Natural History Society's report was momentously persuasive in drawing international attention to the condition of rhinoceros. The Scientific Conference on Resource Conservation and Utilisation of UNESCO held in 1949 urged the Assam government to take 'such steps . . . to protect and preserve the Great One-horned rhinoceros . . . especially in the Kaziranga Sanctuary which should be set aside as an inviolable rhinoceros sanctuary'.[98] A note prepared by Salim Ali, based on the BNHS's report, found a place in the proceedings of this crucial conference. Thus began the efforts to understand and study the rhinoceros.

This was, however, a time in which the idea of the threat of human and domestic animals posing a challenge to the existence of the rhinoceros came to become part of the Kaziranga governance narrative. S. Dillon Ripley and Salim Ali had only broadly hinted at the idea, but years later the idea of an inviolate Kaziranga became prominent. The occasion was the visit of Lee Merriam Talbot in 1959. Talbot was introduced by the *Times of India* as an unusual hunter who visited the park with 'weapons' consisting of a pen and pencil.[99] Writing in the leading conservation journal *Oryx*, Talbot, staff ecologist of the International Union of Conservation of Nature, reported the health danger posed by domestic animals to the rhinoceros. 'The main threat to the rhinos at Kaziranga', Talbot declared, 'is by domestic stock within the sanctuary boundaries'.[100] Talbot further explained that while the government allowed grazing 'on an area one mile deep and 3 miles long inside the edge of the sanctuary', this hardly worked. He lamented how 'the herds of buffalo and cattle are generally grazed free without supervision, it is extremely difficult to enforce those limits'. Forest officials and conservationists alike agreed that this had resulted in occasional deaths of rhinoceros and other fauna due to disease spread by domestic animals. Talbot argued that anthrax and rinderpest were the worst offenders. As the concept of sanitising the park from human influence towards the end of the 20th century, Talbot's idea would come handy for the later conservationists and managers of the park.

Back at home, beginning in the 1950s the Assamese elite and politicians across the ideological spectrum slowly began to extend support to the park. Gee indeed noticed 'signs of an increasing volume of public opinion in Assam . . . more interested in shooting with the camera'. Gee delightfully wrote how 'of late many speeches and broadcasts on Assam's wonderful flora and fauna have emanated from important officials'.[101] As international conservation scientists regularly drew attention to the endangered condition of the rhinoceros in India, Assamese politicians responded by enacting legal mechanisms to ensure the further safety of the animal.[102] In the next decade extraordinary steps were taken to frame legal and executive mechanisms to give a new life to Kaziranga. But since then the local people also extended incredible social and political support to enhance the value of Kaziranga. People began to care about wild animals, which at one time were a nuisance and a cause of concern for raiding their crops. Despite this momentous new journey taken by the rhinoceros, little on the rhinoceros came to be portrayed in the otherwise politically assertive world of Assamese literature.

A game reserve and its eventual transformation to a national park in the 1970s and the associated conservation initiatives along with legal, financial and political programmes by the government helped save the rhino, proving that large animals including the rhino could be saved from hunters and others. By the last quarter of the 20th century the park slowly began to acquire significant brand value and international attention. Naturalists paid regular visits to the Kaziranga and their findings came to be presented in a wide range of global forums. Bill Burrad, the American TV host, made a short documentary on the Kaziranga titled 'From Kaziranga with Love', making it an international tourist destination. In the 1970s travel to the region was still restricted, requiring permits from the government of India. Thus, for instance, in 1970, the *New York Times* reported how the Assam government had made a few modern rest houses and a transport facility for tourists to see the 400 rhinos who were otherwise not hostile to humans. The 1970s witnessed moderate growth in the numbers of Indian and international visitors to the park, pushing it beyond the exclusive interests of the naturalists. The rich, the children of princely rulers and the elite would make occasional visits to the park. This pattern would continue for two more decades.

By this time there was rapid political and economic transformation in Assam. The political instability of the 1980s resulted in the fragile governance of the park. The rhinos in Kaziranga became part of an international network of medicinal trade and poaching of the rhinos became frequent. News of the indiscriminate killing of the rhinos in Kaziranga came to be covered in the international press. During this time of political restlessness, the rhinos of Kaziranga acted as additional armour for the Assamese. In 1982 the government of India proposed to move about 60 of the rhinos from the park to other parts of the country.[103] While the naturalists supported this move as ecologically justifiable, leaders of the Assam movement termed the proposed transfer an attempt by the central government to deprive the Assamese of a tourist attraction. Bhrighu Phukam, the fiery spokesman for Assamese leaders, claimed that any such plans were unacceptable as it was 'a conspiracy to deprive the people of Assam of an object of their pride and deprive the area of tourist revenue.' In February 1982, a delegation of Assamese leaders who had been negotiating with Prime Minister Indira Gandhi on their demand for the identification and deportation of settlers from Bangladesh also requested that she bar the removal of any animals from Assam.

By the 1980s the park became an integral part of Assamese cultural and political power along with its increasing branding as a tourist destination. This undoubtedly enhanced the park's potential to generate economic opportunities. A few Assamese small entrepreneurs, for whom the space for trade and business was very limited, began to invest private capital and by the turn of the century a vibrant tourism business was in place, largely dominated by Assamese small entrepreneurs. An augmentation of its economic value thus benefitted a good number of small traders and peasant families. Rapid changes in the political economy of the park have prompted the birth of new management practices.

In the meantime, an overarching public discourse, shaped by the sense and sensibilities of the Assamese nationalist imagination, had transformed the park into a space where Assam could claim some political exclusivity. The desire to preserve KNP, well pronounced in the Assamese public life, is not essentially driven by a sense of recovery of a natural space, but more specifically due to its embodied economic and cultural value.

Competing rights: people, land and animals

By the turn of the century there were three distinct layers of development around the life of the park: increasing entry of private capital, peasantisation of the floodplains and finally, renewed vulnerability of the rhinoceros to poaching due to its increasing demand in the international medicinal trade. In the latter decades of the 20th century there was an increased proliferation of trade in the rhino horn. Of the international rhino horn markets, Hong Kong, Aden and Yemen were major importers. Between 1949 and 1976, Hong Kong had been the major import market for rhino horns.[104] The Assam government also participated through a legal window in the rhino horn trade during this period. Between 1969 and 1979 the Assam Forest Department sold 210 kg of Indian rhino horn.[105] The increasing vulnerability of the rhinos intensified acrimonious public debates. These debates focused essentially on the competing rights of capital, animals and the peasant population. One line of argument strongly suggests that human influences in the park are now widely believed to be a dangerous evil. Public debate equally distances itself from the interconnected issues of nature and the human landscape, which shaped the past of Kaziranga.

With an increased animal population inside the park, the park authorities were forced to seek territorial expansion. The direction of this territorial expansion was based on a complicated understanding of ecology, bureaucratic conveniences and local political negotiations. While making this new choice for territorial expansion, the park authorities repeated what their imperial predecessors had done: when the park was transformed into a state-owned space, very little was done to investigate and understand the nature of local practices with regard to the use and exploitation of natural resources. The local population's ideas about ecology, land and water use patterns and grazing practices were seldom recognised. The colonial government, while creating game reserves, denied rights to the local people, and this became the standard practice for successive governments.

As the process for land acquisition began, the ghost of the unfinished land settlement programme of the British colonial and post-independent era and the renewed process of peasantisation of 1970s and thereafter had come to haunt the government and the park authorities. For instance, those landholders, whose record of rights was documented in the official land registry, refused to cede their land. Their refusal to relinquish their rights over lands came to be seen by the government as a threat to the idea of conservation. For those whose land tenure rights were never recorded in the standard official land registry and remained fuzzy, displacement meant no compensation. These villagers demanded that their land tenures be first secured before the reclamation process began. While refusing to be rehabilitated or displaced and lamenting the loss of their access to the commons, they contested the undisturbed privilege of private capital and their profit-making activities.

The government's dilemma has emerged out of a range of complexities. First, after 1947 the mechanism and principles of land governance in Assam had never been freed from the ideological arithmetic of colonial bureaucracy. The governing technology of land in colonial Assam was driven by the ideas of ruthless exploitation of peasant labour and expansion of the agrarian frontier. Second, large stretches of 'unclassed state forests', an important component of the colonial forestry in Assam, remained over the years an ill-defined and fluid category. With the onset of the electoral politics since 1930s, after the promulgation of the Government of India Act of 1935, Assamese politicians and the revenue bureaucracy regularly negotiated with the unclassed state forests to meet the rising demands of the agrarian communities for land. Many of them might have lost their lands to river erosion or become victims of the credit-mortgage nexus. As the 20th century progressed and as this category of forests lost its economic importance for the state forestry programme, the Forest Department began to disown and abdicate their ownership; these spaces became new geographies of peasantisation in Assam. In the 1970s, agrarian expansion into these legally forested spaces reached a significant momentum. The Forest Department feebly defended its territorial jurisdiction but this could not stop the march toward peasantisation. By 1980, armed with the Forest (Conservation) Act of 1980, the Forest Department reasserted its proprietorship over these unclassed state forests.

The public debates that shape the modern journey of Kaziranga hardly look at its complex evolution and the larger world of governance of nature in Assam. The present war of words is practically a battle against Assam's agrarian past. Legal battles are fought using 19th-century land revenue and forest laws, which were written to maximise political and economic benefits to the British Empire but failed to resolve critical issues. These laws are not equipped to accommodate the worries of the ecologically devastated Assamese peasantry and that of the Assam's crucial floodplain's fluid environment. Over the years KNP has become an abstract environment, ignoring the vibrant everyday agrarian life around the park. The complex legal and conservation debates also refuse to take cognizance of the domesticated agrarian landscape. But this chapter argues that such an idea of an impeccable natural space, free from all human interaction and influences, will find little root in environmental pasts.

Notes

1 'The politics of cleaning up Kaziranga', *Indian Express*, 27 September 2016.
2 http://ghconline.gov.in/Judgment/PIL662012.pdf [accessed 20 October 2017].
3 Ibid.
4 'Rhino protection a poll issue in Assam but only symbolically', *Hindustan Times*, 7 April 2016.
5 For an insight into the ecology and habitat of the great one-horned rhinoceros, see, E. Dinerstein, *Return of the Unicorns: Natural History and Conservation of the Greater-One Horned Rhinoceros*, New York: Columbia University Press, 2003.
6 P. C. Choudhury (ed.), *Hastividyarnava*, Guwahati: Assam Publication Board, 1976.
7 T. C. Bhattacharjee (ed.), *Ghora Nidan*, Shillong: Assam Government Press, 1932. Bhattacharjee emphatically noted that this work was 'a product of experience acquired independently in Assam with reference to horses trained and domiciled in the land'.
8 For details see Maheswar Neog, *Socio-Political Events in Assam Leading to the Militancy of the Māyāmariyā Vaiṣṇavas*, Calcutta: K.P. Bagchi, 1982.
9 See, Ambika Aiyadurai, '"Tigers are our brothers": Understanding human-nature relations in the Mishmi Hills, Northeast India', *Conservation and Society*, vol. 14, no. 4, 2016, pp. 305–316.
10 Arupjyoti Saikia, 'The Kaziranga National Park: Dynamics of social and political history', *Conservation and Society*, vol. 7, no. 2, 2009, pp. 113–129.
11 Ibid.
12 R. S. S. Baden-Powell, *Pig Sticking and Hog Hunting: A Complete Account for Sports Men and Others*, London: Harrison & Sons, 1889, p. 145.
13 John Butler, *Travels and Adventures in the Province of Assam, During a Residence of Fourteen Years*, London: Smith, Elder and Co., 1856.
14 Ibid.
15 T. T. Cooper, *The Mishmee Hills*, London: Henry S. King, 1873, p. 63.
16 John Butler, *A Sketch of Assam: With Some Account of the Hill Tribes*, London: Smith, Elder and Co., 1847, p. 27.
17 A. H. Meysey Thompson, 'A trip to India', *The Leeds Mercury*, 2 February 1899.

18 Vivek Menon, *Under Siege: Poaching and Protection of Greater One-Horned Rhinoceroses in India*, Cambridge: Traffic International, 1996, p. 2.
19 L. C. Rookmaaker, *The Rhinoceros in Captivity: A List of 2439 Rhinoceroses Kept from Roman times to 1994*, The Hague: SPB Academic Publishing, 1998, pp. 78–79.
20 Mahesh Rangarajan, *India's Wildlife History: An Introduction*, New Delhi: Oxford University Press, 2005, p. 14; Edwin T. Atkinson, *Statistical, Descriptive and Historical Account of the North-Western Provinces of India*, vol. 4, Part I: Agra Division, Allahabad: North-Western Provinces Government Press, p. xiii.
21 See, for instance, Thomas Claverhill Jerdon, *The Mammals of India: A Natural History of All the Animals Known to Inhabit Continental India*, Roorkee: Thompson College Press, 1867.
22 John McClelland, 'A list of Mammalia and birds collected in Assam', in *Annals and Magazine of Natural History: Zoology, Botany, and Geology*, vol. 6, London: R. and J. E. Taylor, 1841, pp. 366–374.
23 'Notes of a pleasure tour in India: No VIII-Assam', *The Glasgow Herald*, 17 August 1885.
24 *Royal Cornwall Gazette*, 10 May 1844.
25 Martin Montgomery, *The History, Antiquity, Topography, and Statistics of Eastern India, Vol. III: Puraniya, Ranggopoor, and Assam*, London: W.H. Allen & Co., 1838, p. 574.
26 George M. Barker, *A Tea Planter's Life in Assam*, Calcutta: Thacker, Spink & Co., 1884, p. 211.
27 Butler, *A Sketch of Assam*, p. 29.
28 Ibid.
29 Ibid.
30 Nripendra Narayan Bhupa and Maharajah of Cooch Behar, *Thirty-Seven Years of Big Game Shooting in Cooch Behar the Duars, and Assam: A Rough Diary*, Bombay: Times Press, 1908.
31 *The Hampshire Advertiser Country Newspaper*, 9 April 1999.
32 'The colonial and India exhibition', *Nottinghamshire Guardian*, 28 May 1886.
33 *The Lancashire Daily Post*, 29 December 1902.
34 Gray noted that the rhino horn fetched as much as Rs 80 to Rs 100 per seer. J. Errol Gray, 'Diary of a journey to the Borkhamti and the sources of the Irrawaddy, 1893', in *Government of India Foreign Affairs Proceedings*, May 1894, Nos. 7–15 (Elwin, Verrier, ed.), 1959. *India's North-East frontier in the nineteenth century. London*, London: Oxford University Press, p. 423.
35 J. H. Baldwin, *The Large and Small Game of Bengal and the North-Western Provinces of India*, 2nd ed., London: Henry S. King and Co., 1877.
36 Quoted in E. Blyth, 'Asiatic rhinoceroses', *The Zoologist*, June 1872, vol. 7, second series, p. 3108.
37 W. L. Scalater, 'Rhinoceros', *Catalogue of Mammalia in the Indian Museum*, vol. 2, 1891, p. 202.
38 Pollock, F.T., 1879, *Sports in British Burma, Assam, and the Cossyah and Jyntiah Hills, with notes of sport in the hilly districts of the northern division, Madras Presidency, indicating the best localities in those countries for sport, with natural history notes, illustrations of the people, scenery, and game, together with maps to guide the traveller or sportsman, and hints on weapons, fishing-tackle, etc., best suited for killing game met with in those provinces*, vol.1, London: Chapman and Hill. p. 94.
39 Earl of Ronaldshay. *The Life of Lord Curzon, Being the Authorized Biography of George Nathaniel Marquess Curzon of Kedleston, K.G.*, vol. 2, London: Ernest Benn Ltd., 1928, p. 168.

40 Ibid.
41 'The game preserves of Assam', *Forest and Stream*, 1910, pp. 1036–1037.
42 R. Mukherjee (ed.), 'Game preservation in India', *Lord Curzon*.
43 'The game preserves of Assam'.
44 M. Gordon Stables, *Tea, the Drink of Pleasure and Health*, London: Field and Tuer, Leadenhall Press, 1883.
45 *Sevenoaks Chronicle and Kentish Advertiser*, April 21, 1905.
46 'Adventure of viceroy's military secretary', *The Scotsman*, 24 February 1909.
47 Jack B. Watson, *Empire to Commonwealth, 1919 to 1970*, London: J. M. Dent & Sons, 1971, p. 1.
48 *Sevrnoaks Chronicle and Kentish Advertiser*, 13 February 1903.
49 *Nottingham Evening Post*, 15 September 1933.
50 E. P. Stebbing, 'Big-game shooting in India', *The Scotsman*, 24 December 1921.
51 Francis Henry Skrine, 'Bengal: The land and its people', *Dundee Evening Telegraph*, 28 January 1902.
52 'Vanishing creatures of the wild: Unless protected, many species of animals will become extinct', *Edinburgh Evening News*, 5 July 1932.
53 Gee, E.P. 1952. The great Indian one-horned Rhinoceros. *Oryx* 1(5): 224–227.
54 Vivek Menon, 'A problem with pachyderms', *Seminar*, no. 466, 1998, p. 47.
55 A. J. W. Milroy, 'The preservation of wild life in India, no. 3 Assam', *Journal of the Bombay Natural History Society*, vol. 37, no. 1, Supplement, 1934, pp. 97–104.
56 Gee, 'The Rhino in Kaziranga', p. 214.
57 Milroy, 'The preservation of wild life in India', p. 99.
58 This and the following sections, if not mentioned otherwise, are based on previous published work, Saikia, 'The Kaziranga National Park'.
59 Letter from J. C. Arbuthnott, Officiating Commissioner of the Assam Valley Districts to the Secretary to Chief Commissioner of Assam, 4 November 1902, no. 2409 G, Gauhati in Assam Secretariat Proceedings, Revenue A, September 1905, nos. 75–134 (Assam State Archives, hereafter ASA).
60 Letter from Secretary to the Chief Commissioner of Assam to J. C. Arbuthnott, Deputy Secretary, Sylhet, no. 2160 Misc, 9628 G Shillong, 18 December 1902, no. 76 in Assam Secretariat Proceedings, Revenue A, September 1905, nos. 75–134 (ASA).
61 Colonisation of Wastelands in Assam, Revenue A, November 1898, File nos. 128–138 (ASA).
62 Quoted in Letter from J.C. Arbuthnott to Secretary to the Chief Commissioner of Assam, 28 August 1903, Jowai, no. 77 in Assam Secretariat Proceedings, Revenue A, September 1905, nos. 75–134 (ASA).
63 Ibid.
64 Letter from F. J. Monahan, Secretary to the Chief Commissioner of Assam to the Commissioner of the Assam Valley Districts, no. 12 Forests, 1283 R, Shillong, 15 March 1904, in Assam Secretariat Proceedings, Revenue A, September 1905, nos. 75–134 (ASA).
65 Letter from J. Donald, Deputy Commissioner of Sibsagar to the Commissioner of the Assam Valley Districts, no. 409 R, Sibsagar, 4 June 1904 in Assam Secretariat Proceedings, Revenue A, September 1905, nos. 75–134 (ASA).
66 Ibid.
67 For a fuller understanding of the American Bison story, see Andrew C. Isenberg, *The Destruction of the Bison: An Environmental History, 1750–1920*, Cambridge: Cambridge University Press, 2000.
68 Saikia, 'The Kaziranga National Park'.
69 H. Blochmann, Koch Bihar, Koch Hajo, and Asam in the 16th and 17th centuries, according to the Akbarnamah, the Padishahnamah, and the Fathiyah i Ibriyah, Journal of Asiatic Society of Bengal, 1872, vol. 41, no. 1, p. 76

70 Letter from J. Donald, Deputy Commissioner of Sibsagar to the Commissioner of the Assam Valley Districts, no. 409 R, Sibsagar, 4 June 1904 in Assam Secretariat Proceedings, Revenue A, September 1905, nos. 75–134 (ASA).
71 Saikia, 'The Kaziranga National Park'.
72 'Lord Minto's tour', *The Times*, 24 February 1909, p. 5.
73 *The Times*, 6 September 1912, p. 3.
74 Gee, 'The rhino of Kaziranga', p. 216.
75 Rangarajan, *India's Wildlife History*, p. 57.
76 Lee Meeriam Talbot, 'A report on some animals of the middle east and southern Asia which are threatened with extermination', *Oryx*, vol. 5, nos. 4–5, May 1960, pp. 155–293, 190.
77 Ibid.
78 Gee, 'The rhino of Kaziranga', p. 216.
79 Assam State Archives, 'Petition to keep from miles of the Kaziranga Reserve open from the boundary line near the villages', Assam Secretariat Revenue proceedings, Revenue and Forests B, nos. 120–128, 1925 (ASA).
80 E. P. Gee, 'The most famous Rhino', *Natural History*, vol. 63, 1954, pp. 366–369.
81 Milroy, 'The preservation of wild life in India'.
82 Ibid., p. 98.
83 H. S. Wood, *Glimpses of the Wild: An Observer's Notes and Anecdotes on the Wild Life of Assam*, London: H.F. & G. Witherby, 1936, p. 115 (Wood was a civil surgeon in Assam).
84 For an understanding of the growth of Nepali milk producers in the first decades of the 20th century, see Lopita Nath, 'Migration, insecurity and identity: The Nepali dairymen in India's Northeast', *Asian Ethnicity*, vol. 7, no. 2, 2006, pp. 129–148.
85 For Bengal jute industry, see Tariq Omar Ali, *The Envelope of Global Trade: The Political Economy and Intellectual History of Jute in the Bengal Delta, 1850s to 1950s*, unpublished doctoral dissertation, Harvard University, 2012.
86 Arupjyoti Saikia, 'Jute in the Brahmaputra Valley: The making of flood control in twentieth-century Assam', *Modern Asian Studies*, vol. 49, no. 5, September 2015, pp. 1405–1441.
87 For an overview of the Assam tea plantations, see Rana P. Behal, *A Century of Servitude: Political Economy of Tea Plantations in Colonial Assam*, New Delhi: Tulika, 2014.
88 Arupjyoti Saikia, *Forests and Ecological History of Assam*, New Delhi: Oxford University Press, 2011.
89 Saikia, 'Jute in the Brahmaputra Valley'.
90 Arupjyoti Saikia, *Elephants, Zamindars and State: History of Contested Hunting Rights in Western Assam*, seminar paper presented at SOAS 2016 Elephant Conference, Indian Institute of Science, April, 2016.
91 'Seclusion for rhino', *Times of India*, 17 April 1952. A rhino was gifted to Periyar Sanctuary in Kerala in 1952. *Times of India*, 21 January 1957.
92 *Times of India*, 26 April 1967.
93 E. P. Gee, 'Wild life reserves in India: Assam', *Journal of the Bombay Natural History Society*, vol. 49, no. 1, April 1950, pp. 81–89.
94 'Survey of wild life in Assam: Experts' report', *Times of India*, 29 July 1949.
95 Gee, 'Wild life reserves in India'.
96 'Wild life', *The Times of India*, 6 August 1949.
97 In 1958 the *New York Times* reported that two people were killed by rhinos coming out of the park. Such stray animals were guarded back to the sanctuary without harming them. 'Indian rhinoceros kills two', *The New York Times*, 10 November 1957, p. 137.

98 Gee, 'Wild life reserves in India', p. 84.
99 Emile C. Schurmacher, 'He risks his life stalking world's rarest animals', *Times of India*, 12 July 1959.
100 Talbot, 'A report on some animals of the middle east and southern Asia'.
101 Gee, 'Wild life reserves in India', p. 88.
102 Saikia, 'The Kaziranga National Park'.
103 S. Gupta, 'Assam anger to move to shift rhinos', *The Indian Express*, 21 April 1982.
104 I. S. C. Parker and Esmond Bradley Martin, 'Trade in African rhino horn', *Oryx*, vol. XV, no. 2, November 1979, p. 157.
105 Esmond Bradley Martin, *Rhino Exploitation: The Trade in Rhino Products in India, Indonesia, Malaysia, Burma, Japan and South Korea*, Hong Kong: World Wildlife Fund, 1983, pp. 35–36.

References

A Rhino was Gifted to Periyar Sanctuary in Kerala in 1952. (1957, January 21). *Times of India*.

Adventure of Viceroy's Military Secretary. (1909, February 24). *The Scotsman*.

Aiyadurai, Ambika. (2016). Tigers Are Our Brothers: Understanding Human-Nature Relations in the Mishmi Hills, Northeast India. *Conservation and Society*, 14(4), 305–316.

Anonymous. (1910). The Game Preserves of Assam. *Forest and Stream*. July 16, pp. 117–118.

Assam State Archives. (1925). *Petition to Keep from Miles of the Kaziranga Reserve Open from the Boundary Line Near the Villages*. Assam Secretariat Revenue Proceedings, Revenue and Forests B, Nos. 120–128 (ASA).

Atkinson, Edwin T. (1876). *Statistical, Descriptive and Historical Account of the North-Western Provinces of India* (Vol. 4). Part I: Agra Division. Allahabad: North-Western Provinces Government Press.

Baden-Powell, R. S. S. (1889). *Pig Sticking and Hog Hunting: A Complete Account for Sports Men and Others*. London: Harrison & Sons.

Baldwin, J. H. (1877). *The Large and Small Game of Bengal and the North-Western Provinces of India* (2nd ed.). London: Henry S. King and Co.

Barker, George M. (1884). *A Tea Planter's Life in Assam*. Calcutta: Thacker, Spink & Co.

Behal, Rana P. (2014). *A Century of Servitude: Political Economy of Tea Plantations in Colonial Assam*. New Delhi: Tulika.

Bhattacharjee, T. C. (Ed.). (1932). *Ghora Nidan*. Shillong: Assam Government Press.

Bhupa, Nripendra Narayan. (1908). *Maharajah of Cooch Behar, Thirty-Seven Years of Big Game Shooting in Cooch Behar the Duars, and Assam: A Rough Diary*. Bombay: Times Press.

British Association, 'The Preservation of Wild Animals' (1912, September 6), p. 3.

Blyth, E. (1972). Asiatic Rhinoceroses. *The Zoologist*, 7(2), 3108.

Blochmann, H., Koch Bihar, Koch Hajo, and Asam in the 16th and 17th centuries, according to the Akbarnamah, the Padishahnamah, and the Fathiyah i Ibriyah, *Journal of Asiatic Society of Bengal*, 1872, vol. 41, no. 1, p. 49–101.

Butler, John. (1847). *A Sketch of Assam: With Some Account of the Hill Tribes*. London: Smith, Elder & Co.

Butler, John. (1856). *Travels and Adventures in the Province of Assam, During a Residence of Fourteen Years*. London: Smith, Elder and Co.

Choudhury, P. C. (Ed.). (1976). *Hastividyarnava*. Guwahati: Assam Publication Board.

Cooper, T. T. (1873). *The Mishmee Hills*. London: Henry S. King.

Dinerstein, E. (2003). *Return of the Unicorns: Natural History and Conservation of the Greater-One Horned Rhinoceros*. New York: Columbia University Press.

Earl of Ronaldshay. (1928). *The Life of Lord Curzon, Being the Authorized Biography of George Nathaniel Marquess Curzon of Kedleston, K.G.* (Vol. 2). London: Ernest Benn Ltd.

G, Gauhati in Assam Secretariat Proceedings, Revenue A, September 1905, Nos. 75–134 (ASA).

Gee, E. P. (1950). Wild Life Reserves in India: Assam. *Journal of the Bombay Natural History Society*, 49(1), 81–89.

Gee, E. P. (1954). The Most Famous Rhino. *Natural History*, 63, 366–369.

Government of Assam, Colonisation of Wastelands in Assam Secretariat Proceedings, Revenue A, November 1898, File Nos. 128–138 (Assam State Archives, hereafter ASA).

Gray, J. Errol. (1893). Diary of a journey to the Borkhamti and the sources of the Irrawaddy, 1893, in *Government of India Foreign Affairs Proceedings,* May 1894, Nos. 7–15 (Elwin, Verrier, ed.), 1959. *India's North-East frontier in the nineteenth century*. London: Oxford University Press, p. 423.

Greater Britain, Chronicle and Kentish Advertiser (1903, February 13), p. 2.

Gupta, S. (1982, April 21). Assam Anger to Move to Shift Rhinos. *The Indian Express*.

Indian Rhinoceros Kills Two. (1957, November 10). *The New York Times*, p. 137.

Isenberg, Andrew C. (2000). *The Destruction of the Bison: An Environmental History, 1750–1920*. Cambridge: Cambridge University Press.

Jerdon, Thomas Claverhill. (1867). *The Mammals of India: A Natural History of All the Animals Known to Inhabit Continental India*. Roorkee: Thompson College Press.

Letter from J. C. Arbuthnott, Officiating Commissioner of the Assam Valley Districts to the Secretary to Chief Commissioner of Assam, November 4, 1902, No. 2409.

Letter from Secretary to the Chief Commissioner of Assam to J. C. Arbuthnott, Deputy Secretary, Sylhet, No. 2160 Misc, 9628 G, Shillong, December 18, 1902, No. 76 in Assam Secretariat Proceedings, Revenue A, September 1905, Nos. 75–134 (ASA).

Letter from J. C. Arbuthnott to Secretary to the Chief Commissioner of Assam, August 28, 1903, Jowai, No. 77 in Assam Secretariat Proceedings, Revenue A, September 1905, Nos. 75–134 (ASA).

Letter from F. J. Monahan, Secretary to the CC, Assam to the Commissioner of the Assam Valley Districts, No. 12 Forests, 1283 R, Shillong, March 15, 1904 in Assam Secretariat Proceedings, Revenue A, September 1905, Nos. 75–134 (ASA).

Letter from J. Donald, Deputy Commissioner of Sibsagar to the Commissioner of the Assam Valley Districts, No. 409 R, Sibsagar, June 4, 1904 in Assam Secretariat Proceedings, Revenue A, September 1905, Nos. 75–134 (ASA).

Lord Minto's Tour. (1909, February 24). *The Times*, p. 5.

Martin, Esmond Bradley. (1983). *Rhino Exploitation: The Trade in Rhino Products in India, Indonesia, Malaysia, Burma, Japan and South Korea*. Hong Kong: World Wildlife Fund.

McClelland, John. (1841). A List of Mammalia and Birds Collected in Assam. In *Annals and Magazine of Natural History: Zoology, Botany, and Geology* (Vol. 6). London: R. and J. E. Taylor.

Menon, Vivek. (1996). *Under Siege: Poaching and Protection of Greater One-Horned Rhinoceroses in India*. Cambridge: Traffic International.

Menon, Vivek. (1998). *A Problem with Pachyderms*. Seminar, No. 466.

Milroy, A. J. W. (1934). The Preservation of Wild Life in India, No. 3 Assam. *Journal of the Bombay Natural History Society*, 37(1), 97–104.

Montgomery, Martin. (1838). *The History, Antiquity, Topography, and Statistics of Eastern India, Vol. III: Puraniya, Ranggopoor, and Assam*. London: W. H. Allen & Co.

Mukherjee, R. (Ed.). Game Preservation in India. *The Great Speeches of Modern India*, Delhi: Randon House, 2007.

Nath, Lopita. (2006). Migration, Insecurity and Identity: The Nepali Dairymen in India's Northeast. *Asian Ethnicity*, 7(2), 129–148.

Neog, Maheswar. (1982). *Socio-Political Events in Assam Leading to the Militancy of the Māyāmariyā Vaiṣṇavas*. Calcutta: K. P. Bagchi.

Notes of a Pleasure Tour in India: No VIII-Assam. (1885, August 17). *The Glasgow Herald*.

Nottingham Evening Post (1933, September 15).

Parker, I. S. C. and Martin, Esmond Bradley. (1979). Trade in African Rhino Horn. *Oryx*, 15(2), 157.

Prince of Wels's India Tour, Chronicle and Kentish Advertiser (1905, April 21), p. 6.

Rangarajan, Mahesh. (2005). *India's Wildlife History: An Introduction*. New Delhi: Oxford University Press.

Rhino Protection a Poll Issue in Assam but Only Symbolically. (2016, April 7). *Hindustan Times*.

Rookmaaker, L. C. (1998). *The Rhinoceros in Captivity: A List of 2439 Rhinoceroses Kept from Roman Times to 1994*, The Hague: SPB Academic Publishing.

Royal Cornwall Gazette (1844, May 10).

Saikia, Arupjyoti. (2009). The Kaziranga National Park: Dynamics of Social and Political History. *Conservation and Society*, 7(2), 113–129.

Saikia, Arupjyoti. (2011). *Forests and Ecological History of Assam*. New Delhi: Oxford University Press.

Saikia, Arupjyoti. (2015). Jute in the Brahmaputra Valley: The Making of Flood Control in Twentieth-Century Assam. *Modern Asian Studies*, 49(5), 1405–1441.

Saikia, Arupjyoti. (April 2016). *Elephants, Zamindars and State: History of Contested Hunting Rights in Western Assam*, seminar paper presented at SOAS, 2016 Elephant Conference, Indian Institute of Science.

Seclusion for Rhino. (1952, April 17). *Times of India*.

Scalater, W. L. (1891). Rhinoceros. *Catalogue of Mammalia in the Indian Museum*, 2.

Schurmacher, Emile C. (1959, July 12). He Risks His Life Stalking World's Rarest Animals. *Times of India*.

Skrine, Francis Henry. (1902, January 28). Bengal: The Land and Its People. *Dundee Evening Telegraph*.

Stables, M. Gordon. (1883). *Tea, the Drink of Pleasure and Health*. London: Field and Tuer, Leadenhall Press.
Stebbing, E. P. (1921, December 24). Big-Game Shooting in India. *The Scotsman*.
Survey of Wild Life in Assam: Experts' Report. (1949, July 29). *Times of India*.
Talbot, Lee Meeriam. (1960). A Report on Some Animals of the Middle East and Southern Asia Which Are Threatened with Extermination. *Oryx*, 5(4–5), 155–293.
Tariq Omar, Ali. (2012). *The Envelope of Global Trade: The Political Economy and Intellectual History of Jute in the Bengal Delta, 1850s to 1950s* (Unpublished Doctoral Dissertation). Harvard University.
Times of India Assam rhino price increased (1967, April 26).
The Colonial and India Exhibition. (1886, May 28). *Nottinghamshire Guardian*.
The Hampshire Advertiser Country Newspaper (1999, April 9).
The Lancashire Daily Post (1902, December 29).
The Politics of Cleaning Up Kaziranga. (2016, September 27). *Indian Express*.
Thompson, A. H. Meysey. (1899, February 2). A Trip to India. *The Leeds Mercury*.
Vanishing Creatures of the Wild: Unless Protected, Many Species of Animals Will Become Extinct. (1932, July 5). *Edinburgh Evening News*.
Watson, Jack B. (1971). *Empire to Commonwealth – 1919 to 1970*. London: J. M. Dent & Sons.
Wild Life. (1949, August 6). *The Times of India*. http://ghconline.gov.in/Judgment/PIL662012.pdf [accessed October 20, 2017].
Wood, H. S. (1936). *Glimpses of the Wild: An Observer's Notes and Anecdotes on the Wild Life of Assam*. London: H. F. & G. Witherby.

11

ENVIRONMENTAL MOVEMENTS AND THE INDIAN SUPREME COURT

Geetanjoy Sahu

Over the last three decades, the Indian Supreme Court has been approached time and again by various environmental groups to protect and improve the environment and ensure individuals live in a healthy environment. Since the emergence of the Chipko movement in the early 1970s, a number of environmental struggles in India have protested against the current use and management of natural resources in different parts of the country. To name a few: the movement against the hydroelectric power plant in the Silent Valley; the movements against the Tehri Dam, Sardar Sarovar project and Polavaram project; the protest against illegal mining activities in Orissa and in many parts of central and eastern India; the Bishnoi community's campaign to protect sacred trees and animals; the Dongria Kondh tribe's campaign to protect the sacred Niyamgiri Hills in Odisha; the campaign to protect wildlife and biodiversity; the campaign in urban India to control air and water pollution; the traditional fishing community's protests against development projects in coastal areas; Hindu religious groups' protest against the pollution of the river Ganga and the construction of the Sethusamudram shipping canal project; and the campaign to protect the Taj Mahal from pollution. These movements have employed a number of strategies ranging from public activism (mass mobilisation of people, dharana, etc.) to knowledge-based activism (using social media, writing and reporting and research work) to draw the attention of policy makers and the Indian judiciary. While policy makers and electoral politics both at the national and state level have failed to address environmental issues, the Indian judiciary has taken an active interest in addressing environmental injustice across the country.

This chapter explores how and what kinds of environmental values are given priority by the judiciary and the impact of judicial decisions on environmental movements in India through a review of a series of environmental judgments of the Indian Supreme Court from 1980 to 2010. The chapter begins by discussing different environmental values that are emphasised and prioritised by different environmental groups in the country and then examines the decisions of the Indian Supreme Court vis-à-vis the litigation filed by environmental groups and activists. Finally, the chapter analyses the challenges for environmental movements in India in seeking justice through the courtroom.

Environmental discourse in India: an overview

Historian Ramachandra Guha (1991) points out that although the genesis of concern for environmental protection in India can be traced back to the early 20th century, manifesting in the form of protests against the commercialisation of forest resources during the British colonial period, it was only in the 1970s that concern for the environment received the public attention. As with other parts of the world, the green consciousness movement for environmental protection and regulation emerged in India in the 1970s and assumed the form of public appeal in subsequent years. The emergence of a number of environmental struggles, including the Chipko agitation against the commercialisation of forest resources; the movement against the hydroelectric power plant in the Silent Valley; the movements against the Tehri Dam, Sardar Sarovar project and Polavaram project; the protest against illegal mining activities in Orissa and in many parts of central and eastern India; the Bishnoi community's campaign to protect sacred trees and animals; the Dongria Kondh tribe's campaign to protect the sacred Niyamgiri Hills in Odisha; the campaign to protect wildlife and biodiversity; the campaign in urban India to control air and water pollution; the traditional fishing community's protests against development projects in coastal areas; Hindu religious protests against the pollution of the holy river Ganga and the construction of the Sethusamudram shipping canal project; and the campaign to protect the Taj Mahal from pollution, reflect a concern for environmental protection that is motivated by various reasons. While it is undoubtedly true that public awareness of environmental issues has been increasing, it is also a fact that green discourse in India has not produced an ideologically homogeneous strand for the protection of the environment (Sharma 2012). There exist many different shades of green in India, each representing different environmental concerns, values and pressure groups. To better understand the nature of Indian green discourse it is important to examine how the varying ideological stances of different social groups are structured, negotiated and reflected in the public discourse. The environmental groups have taken diverse socio-economic, political and material positions in their emphasis of sustainable use of resources. In order to understand how the Indian Supreme Court addressed the diverse positions of Indian environmental movements, we must first understand what these positions are. The following section gives an overview of the diverse values advocated by environmental movements in India.

The term 'green' in India has different meanings for different parts of society. In this section, I argue that there are different shades of green, i.e., different environmental values, which are framed, constructed and expressed by different groups from their own particular social, cultural, material, religious and political perspectives. One of the major strands of green discourse in India relates to the entitlement of different social groups to natural resources and also to the ways in which their immediate livelihood depends on the use and management of these natural resources. Scholars have long argued that there is a direct and intricate relationship between the use and management of natural resources and the livelihoods of a majority of people and that any change in this relationship might result in a number of environmental struggles (Guha 1991; Baviskar 1995; Guha and Martinez-Alier 1998). A majority of the people in India are largely dependent on natural resources not only for their livelihood but also for their continued existence. Most environmental movements advocate that people who substantially depend on natural resources for their livelihood be able to use as well as control them, and also encourage alternative use of natural resources. Martinez-Alier (2002) calls these movements the environmentalism of the poor, as distinct from Western environmentalism, which he believes is a product of affluence. Like Martinez-Alier (2002) and Ramchandra Guha (1991), many scholars (Narain 2002; Lele 2011; Sethi 1993) have argued that Indian environmentalism is different from Western environmentalism due to its emphasis on social justice and equity in its demand for the sustainable use of resources. Indian environmentalism has also played a critical role in shaping and reshaping the demands of forest dwellers and rural and traditional communities, including tribals and farmers, in the use and management of natural resources. This brand of green discourse emphasises that because a majority of the people in India, especially poor farmers and tribal communities, are dependent on natural resources for their livelihood, they should be readily able to use and manage natural resources on a sustainable basis. Thus, green discourse in India is largely motivated by a material interest in the environment as a source and requirement for livelihood and by a concern for the rights of the poor rather than for the rights of other species or of future generations of humanity (Martinez-Alier 2002).

Another strand of green discourse in India, however, points out that increased concern for the environment in India is not always necessarily for material interest and that it can be for the nonmaterial and intrinsic value of nature. Many scholars argue that human concern for preserving the quality of the environment is directed towards seeking a change in the official policy to ensure a healthy environment and emphasising the fact that human beings are one among many equal species (Krishna 1996; Sethi 1993). They also argue that the struggle to preserve the quality of environment is quite different from the struggle to protect the environment because of concerns for livelihood. First of all, many conservationists contend that nature has an intrinsic value in and of itself, apart from its contribution to human well-being. They maintain that all created things are equal and that they should be considered not as means to an end but rather as ends themselves, having a right to their own existence without human interference. They value biodiversity for its own sake and assign to the rest of the nature an ethical status at least equal to that of human beings. Some even contend that the collective needs of nonhuman species and inanimate objects must take precedence over man's needs and desires. Animals, plant species, rocks, land, water bodies and so forth, are all said to possess intrinsic value in terms of their very existence, irrespective of their relationship to human beings (Kolstad 2006; Madhusudan and Shankar 2003). These ideas have a significant measure of support both from the state and middle-class environmentalists in India in the form of wilderness movements (Guha 1997).

Furthermore, other scholars point out that environmental concern motivated by the desire for a healthy environment comes primarily from the middle class, which has no direct material interest in the environment. This concern stems from the opinion that people have a basic right to live in a healthy environment, which middle-class environmentalists claim is non-negotiable and must be ensured through various policy initiatives, including scientific and technical measures. This kind of middle-class environmental activism has been criticised by social scientists, most notably by the environmental sociologist and writer Amita Baviskar (2011), who sees the closing down of industries in the name of larger public interest as a strategy adopted by middle-class society, along with various state and non-state actors, such as the courts, lawyers, bureaucrats and environmental activists, to decide which environmental values should be prioritised, while ignoring the rights of workers. Urban middle-class environmentalism differs significantly from the environmentalism of the poor, which emphasises a democratic and participatory approach to protecting the environment and using natural resources and supports devolution of power from the state to the community.

This analysis of the various facets of green discourse in India and other related issues suggests that the domain of green is not only about the protection of environment. Rather, greenness takes on different shades with different goals and different methods of achieving those goals. At the same time, as Baviskar (2005) suggests, these diverse values are interconnected. However, a close look at the evolution of Indian environmental policy and the decisions made by various state machineries, starting from the Wild Life Protection Act of 1972 to the Forest Rights Act of 2006, suggests that the recognition and legitimation of any one particular environmental, socio-economic, cultural and political value over others depends on the clout that the group advocating the issue wields with the existing political regime, the strategies that the group uses and most importantly the nature of the prevalent political economy. Keeping in mind the different shades of green, their respective environmental concerns and their importance in the environmental decision-making process, I have explained how the Supreme Court has interpreted and emphasised green values in its judgments and have accordingly analysed its greenness.

Environmental litigation and the Indian Supreme Court

The increasing role of the Supreme Court of India in resolving environmental disputes has of late come to be recognised as an important feature of environmental governance in India. It is worth noting here that not a single week passes without an environmental judgment delivered either by the Supreme Court or high courts somewhere in the country directing the implementing agencies concerned to protect and improve the environment. The reasons underlying this state of affairs appear varied and complex, however, one major factor seems to be the ineffective implementation of the laws concerned. Many legal and social science scholars widely agree that there is governance deficit in the field of environmental protection and improvement. This has prompted environmentalists and the people, as well as non-governmental organisations, to approach the courts, particularly the higher judiciary, to seek suitable remedies. Interestingly, the judiciary has responded in a proactive manner to deal with these different environmental problems.

The environmental judgments of the Indian Supreme Court, however, are not consistent and represent diverse environmental values as advocated by different environmental groups in the country. In this regard, it is worth discussing what kind of green approach the Indian Supreme Court supports and how its green approach has impacted environmental movements in India.

In order to understand the approach of the Supreme Court of India vis-à-vis litigation filed by diverse environmental groups in India, I have examined a selective number of environmental orders and judgments from 1980 to 2010.[1] The cases under review are diverse in nature, ranging from water and air pollution to forest degradation. The review of environmental judgments is useful in identifying the approach of the Indian Supreme Court towards diverse values as advocated by environmental groups in India.

A list of illustrative environmental judgments: an overview

In *Rural Litigation & Entitlement Kendra and Others v. State of Uttar Pradesh and Others*, the petitioner RLEK, a local NGO, pleaded for the cancellation of a large number of leases extended to limestone quarries on the grounds that they had been polluting the environment, causing ecological imbalance and hazards to the health of not only human beings but also of all inanimate and animate things in the process. On the other hand, the respondents including both the state and the limestone quarry units argued that the closing down of limestone quarries would throw the owners out of business in which they had invested large sums of money and also there would be loss to state's revenue.[2] Notwithstanding the resource capacity of the limestone quarry units and the government of Uttar Pradesh, the court in its order observed that

> the closing down of lime-stone quarry units would undoubtedly cause hardship to them, but it is a price that has to be paid for protecting and safeguarding the right of the people to live in healthy environment with minimal disturbance of ecological balance and without avoidable hazard of them and to their cattle, homes and agricultural land and under affectation of air, water and environment.[3]

The order of the Supreme Court in *Indian Council for Enviro-Legal Action v. Union of India* best illustrates how environmental cause and not resource capacity of the parties involved is important. The petitioner, the Indian Council for Enviro-Legal Action, brought this litigation to the Court with a view to reducing the pollution caused by several chemical industrial plants in Bichhri village, Udaipur District of Rajasthan. The respondents, including Hindustan Agro Chemicals Limited and Jyoti Chemicals, were operating heavy industry plants there, producing chemicals such as oleum (concentrated form of sulphuric acid), Single Super Phosphate and 'H' acid. Calling them rogue industries, the Court held that these industries had inflicted untold misery upon the poor, de-spoiling their land, their water sources and their entire environment, all in pursuit of private profit, and that they had failed to comply with the statutory acts for prevention and control of pollution.[4] Accordingly, the Court ordered the closure of all these plants. Likewise, in *Tarun Bharat Sangh, Alwar v. Union of India*, the local NGO challenged the government for giving licence to start mining activities in protected areas and requested that the Court direct the government to cancel the licences of all the mining units. With all the resources at its disposal, the government of Rajasthan could not substantiate its decision, resulting in the cancellation of licences given for mining activities in the protected areas.[5]

Similarly, the Vellore Citizens Welfare Forum (VCWF),[6] a local NGO in the Vellore district of Tamil Nadu, filed a public interest litigation against the large-scale pollution of soil and water caused by a number of tanneries (more than 500 tanneries) and other industries in the state of Tamil Nadu. The tannery industries were given the option to either install common effluent treatment plants (CETP) for a cluster of industries or to set up individual pollution control devices. The Tamil Nadu Pollution Control Board had prescribed standards for the discharge of effluents with the central government offering a substantial subsidy for the construction of CETPs. However, the progress was slow, forcing the court eventually to order the closure of several industries.

In the *M. C. Mehta v. Union of India* (a Delhi industrial relocation case),[7] the petitioner argued that hazardous and noxious industries operating in Delhi were violating the Master Plan for Delhi 2001 approved by the central government under Section 11 A (2) of the Act, 1990, while posing a serious threat to public health in the process. In this case, the public interest litigation filed by Supreme Court advocate M. C. Mehta resulted in the closing down of 168 industrial units and the relocation of more than a thousand industrial units from Delhi to other parts of the country. Most importantly, in the Delhi vehicular case, the Supreme Court had to face strong resistance from different quarters including the industrial lobby and government of Delhi when it directed the conversion of all the buses to compressed natural gas (CNG) and emphasised that citizens should have a healthy environment, while sidelining other arguments of the government which were based on certain scientific and technical reports that supported ultra low sulfur diesel.[8] This case illustrates how environment and health issues become more important than the resource capacity of the parties involved in the litigation.

The court decisions, however, with respect to environmental litigation filed against the infrastructure projects, have not gone in favour of the environmental movements, in spite of the environmental groups and movements making all efforts to prove how the state initiated policy violated the Statutory Acts and Constitutional Provisions for the protection and improvement of environment. For instance, in *Narmada Bachano Andolan v. Union of India and Others*, the petitioner made all the efforts through various strategies including the mobilisation of people, intensive critical debates and discussions on development projects and its alternative, use of media, submission of facts and materials through research activities, highlighting the plight of affected people, particularly the tribal and other disadvantaged groups, but the arguments raised by the petitioner were sidelined by the judiciary and development project has been priortised in a significant manner.[9] In contrast to pro-environment judgments, there are a number of environmental litigations where the court has adopted a defensive approach towards protecting the environment and has deviated from its own principles and precedents. The outcomes of the Tehri dam case, Narmada dam case, and the construction of the thermal power plant at Dahanu Taluk are some of the notable examples in which the claims of the underdogs for environmental protection have been rejected.

Despite these judgments going against the concerns and values raised by the environmental movements in India, there have been a number of other environmental litigations wherein the petitioner invoked legal doctrines in drawing the attention of the Court to the apathy of the state agencies for the protection of environment. In a majority of environmental litigation, the state and administrative agencies have been the accused parties while individuals and organisations were the initiators, but the final outcomes of the Court decisions primarily suggest that resource inequality of the parties involved is not the determining factor in the judicial decisions on environmental cases.

Why did environmental movements succeed in the courtroom?

A significant factor determining judicial decisions in favour of environmental protection and the rights of poor and disadvantaged groups of society was the active and consistent involvement of public interest citizens and lawyers. From 1980 to 2010, a number of public interest citizens, lawyers and NGOs and their networks pressured for protection and improvement of the environment and also drew the attention of judiciary to the disproportionate distribution of environmental goods and burdens in different parts of India. The initiatives of Rural Litigation and Entitlement Kendra (RLEK) in filing environmental litigation for the protection and improvement of environment in Dehradun in the early 1980s were perhaps the first nationally known environmental litigations to address environmental problems in Dehradun.

Though in the beginning, public interest citizens were more concerned about water and air pollution issues, many environmental NGOs subsequently took an active interest in demanding the protection of wildlife and forests and also the rights of people living in and around forests. Public interest litigations are now the primary and sometimes the only strategy employed by environmental activists and NGOs. Vimal Bhai, the founder of MATU Jan Sangathan, believes that litigation is the most important strategy and last resort to get environmental laws implemented in the country.[10]

Describing the crucial role played by public interest citizens and environmental groups, Dr. Avdesh Kaushal of Rural Litigation and Entitlement Kendra[11] says that

> public interest citizens brought three things to the environmental jurisprudence process in India: a fact that in a number of cases affected people from environmental hazards cannot afford and aware about the legal and scientific issues revolving around the environmental problem to appeal to the Court of law; a perspective that recognized the unequal distribution of environmental goods and burdens; and finally, it exposed the failure and lack of will among the implementing agencies discharged with constitutional duties for the protection and improvement of environment.

The proliferation of public interest citizens and environmental NGOs from 1980 to 2010 marked an important shift in the judicial approach towards environmental issues. Since the 1980s, there has been a flurry of environmental litigation filed by public interest citizens and NGOs for the protection of the environment for different purposes. For example, M. C. Mehta, an environmental activist and lawyer, filed a number of public interest litigations for the protection and improvement of the environment in different parts of India. Other environmental NGOs and citizens filing public interest litigations include Indian Council for Enviro-Legal Defence Forum, Vellore Citizens Welfare Forum, Rural Litigation and Entitlement Kendra, Goa Foundation, Bombay Environmental Action Group, Dahanu Taluka Environment Protection Group, Research Foundation for Science Technology and Natural Resources Policy, Environmental Supports Group and social activist Vimal Bhai from MATU Jan Sangathan also deserve appreciation for their continuing actions against the existing state of the Indian environment.

These environmental groups and NGOs were also represented by a group of public interest lawyers who took active and consistent interest in arguing environmental litigation and bringing their experiences of legal and constitutional principles for the protection and improvement of the environment.[12] Given their experience and interest on the environmental jurisprudence of other countries also helped them to bring most of the environmental jurisprudence principles from absolute liability to precautionary principle, polluters pay principle, public trust doctrine, intergenerational equity and sustainable development. These principles helped to shape and reshape environmental jurisprudence process in India and protect the rights of underdogs.

One of the main reasons for the success of underdogs in drawing the attention of the judiciary and getting favourable decisions has been the innovative methods followed by public interest lawyers and citizens. While the petitioners used to provide ground realities of the environmental problem and its impact, the lawyers played an important role in drafting the petition and raising all the legal and constitutional issues relevant to the case. Apart from highlighting the legal and constitutional issues, the lawyers had to do a lot of groundwork by visiting the affected area and understanding the various other dimensions related to the litigation. Making regular spot visits and interacting with affected people helped the lawyers to argue their cases in a comprehensive manner and it paid dividends in the outcome of the case.[13] Referring to the active and progressive role of the Indian judiciary in supporting environmental causes, environmental lawyer Mr. Raj Panjwani had to say that "it was rather the subject matter of the petition and not the personality of the petitioner or the resource capacity of the parties involved in the litigation used to determine the outcome of the case".[14] Other environmental lawyers also of the view that the subject matter of the petition and how one drafts the petition and argues in the Court in bringing all dimensions of the case had a significant impact in influencing the outcome of judicial decisions in environmental cases than merely the resource capacity of the parties involved in the litigation.[15]

Another important factor that contributed to the success of environmental movements in achieving their desired goals in the courtroom also includes the evolution of various environmental laws and policies emphasising the need for protection and improvement of the environment. Since the 1970s India has employed a range of regulatory instruments for protecting and improving its environment. However, environmental laws were being practised more in violation than conformity and a large number of industries were operating without proper safety and pollution control measures. This prompted environmental groups and NGOs to draw the attention of the judiciary for the effective implementation of environmental laws. Interestingly, the Indian judiciary took an active interest to use its judicial review power to ensure the effective implementation of laws by the implementing agencies and directing them to discharge their constitutional duties. Justifying its intervention due to the failure of implementing agencies in complying and enforcing environmental law, the Indian Supreme Court has observed that

> Enactment of a law, but tolerating its infringement, is worse than not enacting a law at all. The continued infringement of law, over a period of time, is made possible by adoption of such means which are best known to the violators of law. Continued tolerance of such violations of law not only renders legal provisions nugatory but such tolerance by the enforcement authorities encourages lawlessness and adoption of means which cannot, or ought not to, be tolerated in any civilised society.[16]

With the government authorities not showing any serious concern about the enforcement of laws, and with development taking place for personal gains at the expense of the environment and with disregard of the mandatory provisions of law, some public-spirited persons have been initiating public interest litigations. The legal position relating to the exercise of jurisdiction by the courts for preventing environmental degradation and thereby seeking to protect the fundamental rights of the citizens, is now well settled by various decisions of the Supreme Court. Though criticisms have been made against the Court for its interference in the affairs of the executive, the Court has made it clear that its primary effort, while dealing with environment-related issues, is to see that the enforcement agencies, whether the state or any other authority, take effective steps for the enforcement of the laws.

In addition to these factors, the most important factor that contributed significantly to achieving the goals of the environmental movements through judicial intervention include the demonstration of active interest and increasing intervention by the Indian Supreme Court to protect and improve the environment since 1980 in the post-emergency socio-economic and political changes in India. Many scholars have argued that in the post-emergency period, the Supreme Court of India has evolved from a positivist court into an activist court (Sathe 2001). However, as Upendra Baxi (2000) points out, the active role of Indian Supreme Court has not, as is generally believed, become suddenly activist during the last three decades. It has taken a long time to acquire its present position and it has had to go through many stresses and strains. The Supreme Court of India started off as a technocratic court in the 1950s but slowly started acquiring more power through constitutional interpretation. Its transformation into an activist court has been gradual and imperceptible. This transformation in the role of the Court has corresponded with the political change that came about in India during the last 50 years.[17]

One important task that was undertaken by the Indian Supreme Court in the post-emergency period was to ensure effective enforcement of environmental law and policy by the implementing agency. However, the active role of the Indian Supreme Court in the post-emergency period is not exclusive to environmental issues, since, for a host of reasons, the active role of the Indian Supreme Court has become decisive in respect of various aspects of governance in India. In fact, its role has become crucial and significant in every sphere of governance, which includes issues such as prisoners' rights, child labour, inmates of various asylums, the right of the poor to education, shelter and other essential amenities, sexual harassment of women in the workplace, corruption in public offices, accountability of public servants and utilisation of public funds for development activities (Bhushan 2004). Nevertheless, its role in environmental protection appears to have assumed greater proportions in comparison to other arenas of governance.

The relaxation of the *locus standi* principle and encouraging petitioners to bring environmental issues to the Supreme Court has been hailed as one of the most important factors for addressing the rights of poor and disadvantaged sections of society affected by environmental pollution. The relaxation of *locus standi* principle has paved the way for the development of a body of environment law through judicial edict. Similarly, the issuing of *suo motu* notices by the Court to the state agencies for preventing and controlling pollution is an innovative method, a deviation from its traditional doctrine of resolving disputes.

It is important to note here that in the formal organisation of the judicial system, only an aggrieved party could go to the Court seeking a remedy. The Supreme Court, then under the leadership of Justice V. R. Krishna Iyer and Justice P. N. Bhagwati, exposed the Courts to public-spirited persons pursuing a public cause and thus ensured access to justice for all, including those who were unable to approach the Court on their own, and allowed any third party to bring their sufferings to the notice of the judiciary. As a result, a number of litigations have been brought to judicial attention, including those related to the violation of prisoners' rights, women's rights, child labour and more specifically protection and improvement of the environment. The innovative methods initiated for resolving environmental disputes have almost entirely dominated the environmental jurisprudence process for more than 30 years and thereby upheld the concerns and issues raised by the underdogs in a significant manner.

Further, the Court's efforts towards ensuring the implementation of its directions have a significant impact at the grassroots level. For example, in the Delhi vehicular pollution case, the Court had taken a considered decision in directing the Delhi government to use not only CNG as an alternative fuel but also appointed a committee to implement its orders by 31 March 2001 at the latest. The other areas of judicial decisions reflecting the concerns of environmental movements include deciding on the quantum of compensation for the damage done to the environment and for the victim, giving an award to a petitioner for taking up the initiative, setting up 'green benches' in different state high courts, applying international principles to environmental protection and involving the petitioner in the implementation process of the Court directions.

Challenges ahead

One of the major challenges for environmental movements in addressing their concerns through judicial intervention has been the inconsistent approach of the judiciary in dealing with environmental cases. An overview of the Indian Supreme Court judgments on infrastructure cases suggests that the Court has moved forward and backward, eventually favouring the development goals of the state that ignore the environmental and human rights concerns of the community in its decisions. It is well-known that in the early years of judicial intervention in environmental protection and improvement in the 1980s, a liberal and sensitive judiciary seriously examined environmental issues by taking different viewpoints into consideration.

If recent judgments are any indication, the Court now seems to be heavily tilted in favour of development and infrastructure projects, sidelining environmental issues and the rights of people dependent on the environment for their livelihood. The general attitude shown by the Supreme Court with respect to such issues has been one of non-interference on the premise that they involve certain technical issues and policy matters which could only be effectively addressed by the expert authorities of the executive branch. In the recent past, judges even observed that if a project was stayed on account of a public interest petition which could be subsequently dismissed, the petitioner be made liable to pay for the damages occasioned by the delay in the project. In the words of the Court, "any interim order which stops the project from proceeding further must reimburse all the cost to the public in case ultimately the litigation started by such an individual or body fails".[18]

Another major challenge has also been the inconsistent approach of environmental groups and petitioners involved in the cases. In the recent past, environmental groups and NGOs have found it difficult to get involved in environmental cases in a consistent manner for various practical reasons. Broadly, environmental groups and activists advance three main reasons why they fail to consistently be involved in the litigation process and also in the post-environmental judgment period, to ensure the implementation of court orders at the grassroots level. First, environmental litigations, apart from legal issues, also involve complex scientific and technical issues and it is not possible to convince the Court against the stronger parties in a sustained manner on the legal, socio-economic and environmental problems around a particular project. Second, most environmental problems are diffused and impact a large number of people. It is therefore practically difficult to address the major concerns of each and every individual and affected parties in the litigation process and produce proper facts, figures and statistics. Finally, though India has witnessed a group of public interest citizens and also a group of public interest lawyers supporting the cause of pollution-affected people, what is missing is the consistent and active support of public interest scientists to the cause and concerns raised by environmental movements in India.[19] Notwithstanding these challenges in the courtroom, it is also important for environmental movements in India to employ multiple strategies and not just resort to the courtroom, as the decision of the Court is final, even if it may not be right. Under such circumstances, environmental movements need to exhaust all options before going to the Court and should not seek the judicial intervention as the first resort to address environmental pollution and degradation in India.

Notes

1 For more details, see Geetanjoy Sahu (2014), *Environmental Jurisprudence and the Supreme Court: Litigation, Interpretation and Implementation*, New Delhi: Orient Black Swan Publication Pvt. Ltd.

2 Rural Litigation & Entitlement Kendra and Others v. State of Uttar Pradesh and Others, AIR 1985 SC 652.

3 Rural Litigation & Entitlement Kendra and Others v. State of Uttar Pradesh and Others, AIR 1985 SC 656.

4 Indian Council for Enviro-Legal Action v. Union of India, AIR 1996 (3) SCC 212.

5 Tarun Bharat Sangh, Alwar v. Union of India and Others, AIR 1992 SC 514.

6 VCWF v. Union of India, AIR 1996 SC 2715.

7 M. C. Mehta v. Union of India, AIR 1997(11) SCC 327.

8 Agarwal, A., Anju Sharma and Anumita Roychowdhury (1996), *Slow Murder: The Deadly Story of Vehicular Pollution in India*, New Delhi: Centre for Science and Environment.

9 Interview with Medha Patkar, leader of Narmada Bachano Andolan, Mumbai. Also, for more details about the judgment, see Narmada Bachao Andolan v. Union of India and Others, AIR 2000 SC 3751.

10 Personal interview with Vimal Bhai, leader of MOTU Organisation, Delhi.
11 Telephone interview with Dr. Avdesh Kaushal of Rural Litigation and Entitlement Kendra, Dehradun.
12 Public interest lawyers like Raj Panjwani, Sanjay Parikh, Prashant Bhushan, Colin Gonsalves, Claude Alvares, M. C. Mehta, Ritwick Dutta and Rahul Chowdhary, to name a few, have represented a number of environmental litigations in India.
13 Personal interview with Ritwick Dutta, environmental lawyer, New Delhi.
14 Personal interview with Raj Panjwani, environment lawyer, Supreme Court of India, New Delhi.
15 This conclusion is derived based on the interaction with environmental lawyers, namely Ritwick Dutta, Prashant Bhushan, Sanjay Parikh, Raj Panjwani and T. Mohan Rao.
16 Indian Council for Enviro-Legal Action v. Union of India, AIR 1996 (5) SCC 282–283.
17 Das, G. (2001), 'The Supreme Court: An Overview', in B. N. Kripal, A. Desai, G. Subramanium, R. Dhavan and R. Ramachandran, eds. *Supreme But Not Infallible*, New Delhi: Oxford University Press.
18 Ranauk International v. IVR Construction and Others, AIR 1998 (6) SCALE 456.
19 This inference is made based on my discussion with public interest citizens and environmental groups, namely Vellore Citizens Welfare Forum, Rural Litigation Entitlement Kendra, Goa Foundation, Bombay Environmental Action Group, Medha Patkar, Vimal Bhai, Rajendra Singh, Debi Goenka and many others.

References

Baviskar, A. (1995), *In the Belly of the River: Tribal Conflicts Over Development in the Narmada Valley*, New Delhi: Oxford University Press.

Baviskar, A. (2005), 'Red in Tooth and Claw? Looking for Class Struggles in Nature,' in Raka Ray and Mary Fainsod (eds.), Katzenstein: *Social Movements in India: Poverty, Power, and Politics,* New York, USA: Rowman & Littlefield Publishers, Inc.

Baviskar, A. (2011), 'Cows, Cars and Cycle-Rickshaws: Bourgeois Environmentalism and the Battle for Delhi's Streets,' in Amita Baviskar and Raka Ray (eds.), *Elite and Everyman: The Culture Politics of the Indian Middle Class*, New Delhi: Routledge.

Baxi, U. (2000), 'The Avatars of Indian Judicial Activism: Explorations in the Geographies of [In] Justice,' in K. Verma (ed.), *Fifty Years of Supreme Court of India: Its Grasp and Reach*, New Delhi: Oxford University Press.

Bhushan, Prashant (2004), Supreme Court and PIL, *Economic and Political Weekly*, XXXIX(18), 1770–1774.

Guha, R. (1991), *Environmentalism: A Global History*, New York: Oxford University Press.

Guha, R. (1997), 'The Authoritarian Biologist and the Arrogance of Anti-Humanism in the Third World,' *Ecologist*, Vol. 27, pp. 14–20.

Guha, R. and J. Martinez-Alier (1998), *Varieties of Environmentalism: Essays North and South*, New Delhi: Oxford University Press.

Kolstad, D. Charles (2006), *Environmental Economics*, New Delhi: Oxford University Press.

Krishna, S. (1996), *Environmental Politics*, New Delhi: Sage Publications.

Lele, S. (2011), 'Climate Change and the Indian Environmental Movement,' in Navroz K. Dubash (ed.), *Handbook of Climate Change and India: Development, Politics and Governance*, New Delhi: Oxford University Press.

Madhusudan, M.D. and T.R. Raman Shankar (2003), 'Conservation as if Biological Diversity Matters: Preservation Versus Sustainable Use in India,' in *Conservation and Society*, New Delhi: Sage Publications.

Martinez-Alier, J. (2002), *The Environmentalism of Poor: A Study of Ecological Conflicts and Valuation*, London, UK: Edward Elgar Publishing Limited.

Narain, S. (2002), 'Changing Environmentalism: A Symposium on the Changing Contours of Indian Environmentalism,' *Seminar Magazine*, August 5, pp. 15–20.

Sahu, G. (2014), *Environmental Jurisprudence and the Supreme Court: Litigation, Interpretation and Implementation*, New Delhi: Orient Black Swan Publication Pvt. Ltd.

Sathe, S.P. (2001), *Judicial Activism in India*, New Delhi: Oxford University Press.

Sethi, H. (1993), 'Survival and Democracy: Ecological Struggles in India,' in P. Wignaraja (ed.), *New Social Movements in South Asia: Empowering the People*, New Delhi: Vistaar Publications.

Sharma, M. (2012), *Green and Saffron*, Ranikhet: Permanent Black.

12

WISE SAYINGS FROM AN 'ECOSYSTEM' COMMUNITY

Reflections from a search for challenging neoliberal worldviews on nature

John Kurien

The greatest sense of estrangement we face today as humans is our increasing lack of connectedness with Nature as a whole. Though the resources and processes in Nature provide for our benefit and well-being, we are being increasingly alienated from them. Despite this, we contend that we can control and dominate these processes and resources, and if need be, can replace them with products of human ingenuity and technology. This mentality is in large part the result of our minds being commandeered by the dominant neoliberal ideology of our times.

The neoliberal ideology commodifies the resources and processes of Nature. It is an ideology which prices the sea and assigns monetary value to the pleasure we get from watching the waves lapping the beach sands. It alters the language of discourse. The sea becomes capital; the waves perform services and the sandy beach is considered green infrastructure. All three can then be converted into 'markets' to persuade people, who otherwise obstinately see no monetary value in them, to now believe they are all worthy of utilisation for humans. This is neoliberalism's way of preserving Nature.

Merely putting a price on Nature and not consciously changing the way we perceive and respond to Nature will be at the peril of our very existence on this planet. Nature will rejuvenate. Humans will perish.

Undoubtedly, many of us attempt small beginnings to live in greater sync with Nature. We search for good examples to follow. We read the numerous declarations and manifestos put out at the increasing number of conferences dealing with our common future on this planet. We seek inspiration, wisdom and lessons.

My own search for perspectives on these issues has not come from conscious academic investigations or new innovative approaches to Nature but rather from the many wise sayings of a group of 'ecosystem people'[1] with whom I have worked – ordinary fisherfolk around the world, who relate intimately with Nature for their livelihood, by their interactions with the dynamic aquatic ecosystems of our planet.

Let me share my reflections from this search.

Encounters and wise sayings

I start by recounting four encounters – from the many I have had over the last four decades with fisherpeople – to put in context a sample of their wise sayings.

I describe them in some detail as these encounters have been definitive moments of great learning and unlearning for me. They helped me reframe the way I perceive resources and processes in Nature as well as the interactions between them.

In our world today, we are tutored to look only at the materiality of Nature and treat it as an open tap of resources and a sink for our wastes. Some of the perspectives from fisherpeople provide a radical challenge to this mindset. They also provide insights for us to offer an agenda of resistance to the neoliberal perspective and prevent us from being completely co-opted to that worldview.

Indomitable faith in the sea

The first encounter goes back four decades to the very first days of my tryst with a fishing community in Kerala. I had just graduated from an elite business school and after turning my back on a job in industry I decided to work with a group of fishers who had set up a fish marketing cooperative with the help of a dedicated team of community organisers.

It was a late humid summer evening in 1973. With the sea as the backdrop, I was amidst the *kattumarams*[2] and nets which lay drying on the beach. Seated in front of me were a group of fishers – enthusiastic members of the cooperative. They were my captive audience, keen to listen to my ideas of how together we could better streamline the workings of their organisation.

The huge fluctuation in their daily incomes, caused by risk and uncertainty associated with fish harvest from the sea, was my topic of the day. Sometimes they labour in vain. On some other days they return with a bumper harvest yielding substantial earnings. There is no such thing as an 'average earning' in marine fishing. Given this situation, I was entreating them to save for the 'rainy day'. A state-owned bank had opened a branch in the neighbouring village and I was avidly proposing that they open bank accounts to save their 'surplus incomes' from bumper harvests.

They listened patiently. And enthusiastically too, I thought!

After my *bhashan* (talk) was over, I asked if anyone had doubts. There was a hesitant silence. The fishers exchanged glances with some egging the person sitting next to them to ask if they had doubts. Soon one of the older fishers – Pathrose Gomez – made a reluctant expression of his need to interject. I encouraged him, saying no question is a dumb question.

He first made a V-sign with his right forefingers, brought it to his mouth and spat out the betel-leaf he was chewing. He then raised a rhetorical question which has stayed with me since. It was my first lesson in unlearning.

"I am not sure I fully understood everything you said, but I have a doubt," interjected Pathrose. Gesticulating with his thumb pointed to the sea behind he said, "Are you suggesting that this which forms the terrain of our daily labour, may dry up tomorrow?"

I was taken completely off-guard by his rhetorical query. It was a totally different worldview. Not one which was obsessed or overtly anxious with planning for the future. Not one which was keen to be fully integrated into a new system which they did not comprehend. When they had at close proximity a vast ecosystem which, despite its vagaries, gave freely of its resources and sustained them over generations, why worry about saving money for tomorrow?

Though I was only 23 years of age, given my middle-class background and education, I was so supremely confident that I knew a lot about the ways of the world which I could share with the fishers in my effort to bring a modicum of financial discipline and economic rationale to their lives.

This doubt of the elderly fisher stood my understanding of the world on its head. I had not the faintest clue that it was their total connectedness with *Kadalamma* (Mother Sea) and the fervent and simple faith that she will always provide, which was the basis of the fishers' carefree attitude about the 'rainy day' and their hesitation to 'save' for future eventualities.

Intertwined futures

The second occasion was a decade later.

Joyachen Antony was an undisputed leader of the small-scale fishers in Kerala. He was instrumental in uniting them. Without using religion and caste identities, he organised them along class lines. In 1983 he began a *sathyagraha* (fast) against the unwillingness of the government to control the incessant and destructive trawling boats which were destroying the marine ecosystem and affecting the livelihoods of small fishers. Backing his actions were thousands of fishers who had recently mobilised themselves under the banner of the Kerala Swathantra Matsya Thozhilali Federation (KSMTF). But there were two important unspoken dimensions to the *sathyagraha*. Firstly, Joyachen belonged to an 'outlier' community which was left out of the world-famous Kerala Model of development[3] and secondly, the powerful KSMTF had no political party affiliation – both anomalies which were not missed by keen observers of the socio-political context of the state of Kerala.

However, the bigger anomaly, which had a more pervasive impact in the larger context of the environment–development debate in India, pertained to Joyachen's insightful perception of the inextricable and intertwined future of a declining fish stock and his own community's future.

One day when visiting his *sathyagraha pandal* (tent) in front of the government secretariat I asked him to summarise the purpose of the KSMTF struggle. His answer to me was spontaneous and highly instructive: "This struggle is for the future: ours and that of the fish" he said confidently.

Only ecosystem communities, who depend on the environment and its natural resources for their life and livelihood, can see this inextricable link. We middle-class, social scientists and activists – biosphere people – merely fight for saving the environment and the resources per se. Our livelihoods do not depend on it. But perhaps soon our very lives will!

Three decades ago this insight from Joyachen was an eye-opener for me. Being a staunch supporter of the fishworkers' movement, I thought that both Joyachen and myself were fighting the same battle, but from different standpoints. But after what he stated to me that day, I realised how foolish and naïve I was to hold this view.

For Joyachen it was a fight for the combined future – of his community, the living resource and the marine environment. He was willing to give his life for it. His statement was an expression of 'empirical subjectivity' – a shared condition of deep feeling combined with concrete experience of being a real active fishworker and a community leader.

My concern was at best an honest and fervent conviction of a supporter of the fishworkers – it did not affect my livelihood or my community. It was a third-person perspective of an 'objective reality' which I had slowly got to know through my interactions with fishers.

But as fate should have it, Joyachen was snatched away from us seven years after this event by the very sea whose future, peace and tranquillity he had struggled for.

Freedom to fish

A third occasion of learning and unlearning came to me in Cambodia in 2005.

Cambodia was a country with a torturous and violent past. Just a reminder of the 'Killing Fields' of Pol Pot is adequate to revive the memories of the horrendous suffering of the Khmer people. In his effort to create a 'primitive communist society' as many as two million Cambodians were uprooted from their homes and perished due to starvation and torture.

At the heart of Cambodia is the highly productive lake Tonle Sap which teems with fish. These waters were cordoned off by huge areas of bamboo fencing and nets called fishing lots which were controlled by rich, well-connected and influential individuals who took possession of these areas at government sponsored auctions. In 2000, the government earned as much as USD 2 million from these royalties. There had been a long history of frequently violent conflict between the lot owners and several hundreds of small-scale fishers who lived on the fringes of the lake over the issue of access to fish. Village folk were being harassed, harmed and even killed when accessing even those areas assigned to them by law.

Sensing a political bonanza for his unstable government, in 2000, the prime minister of Cambodia took the initiative of appropriating the individual property rights of a few hundred individual owners who had the exclusive rights to almost 50 per cent of the Tonle Sap. He ordered that these areas be converted to realms where the riparian communities could have the full freedom of exclusive community rights to fish by forming community fisheries organisations. But they would only be allowed to fish using very simple fishing nets and traps. This action of the prime minister later came to be called 'the First Fishery Reform'.[4]

I was visiting Cambodia for the first time in 2005 for a short stint in its inland fisheries research institute to help in reframing its research priorities in the context of the interesting and radical fishery reforms which had been put in place by the government. My first request at the institute was to go to a village where the community had created a community fisheries organisation and try to get first-hand from the community what they perceived to be the benefits they have received five years after the reforms.

A two-hour drive from Phnom Penh got us to the village of Kampong Tralach Leu. I was accompanied by a researcher from the institute and a fisheries officer who was to be my translator.

A small gathering of about 30 persons had arrived at the pagoda where we had agreed to meet. There was a mutual exchange of introductions between us and the chief of the community fisheries organisation, the village headman and other men, women and children who were members of the organisation. On one of the pillars of the hall, a map which designated the official area of the community fisheries organisation was prominently displayed.

After the initial pleasantries, I took the opportunity to eagerly hear from the members about their experience of working together and the benefits they had obtained from access to the resources.

Pointing to the map, I posed my question thus, "Having taken collective control over this fairly large area, with considerable fishery and other resources, what concrete benefits do you have today compared to the years when you were denied access to all this by the lot owners?" A rather typical question from a keen researcher!

There was a stony silence.

I interjected to clarify, allowing more time for translations: "Maybe you now have more fish for consumption; perhaps you feel the family is better fed; maybe you have a little more cash income from sale of the surplus from your daily harvest; maybe you use this money to buy small assets – like a cycle, a TV set; maybe now when someone is sick you have the money to buy medicines; perhaps it is not any more a problem to buy text books for the children; maybe . . . what do you consider are the big benefits?"

Still no one answered. I was beginning to think the silence resulted from the memory of the Pol Pot days – suspicion of outsiders!

Then rather abruptly Kim Soeun, the village headman stood up and spoke in a very passionate manner. One word which he repeated several times in the end of his long intervention sounded like "*Sir-i-pee-up*", "*Sir-i-pee-up*".

When I got the translation, I was astonished at his gentle rebuke to the social scientist in me! Apparently, he said: "Sir, you have mentioned several benefits and to some degree we have enjoyed a bit of each of them. We certainly eat more fish now. We do get a little more cash income. Children go to school more regularly and so on. But you did not mention the main benefit we have obtained from the reforms. Freedom! Freedom! [*Sir-i-pee-up*] Freedom to access resources of nature around us without fear. Everything else you mentioned arises from this freedom."

I was completely taken aback. Did this man read Amartya Sen? Or perhaps more appropriately, did Amartya Sen get his inspiration from people like this?

I realised that as social researchers, more often than not, we very mistakenly think that the poor perceive benefits only in material terms of the market. We assume arrogantly that other attributes like freedom, dignity and self-respect have a much lower priority in their lives. How grossly we underestimate the deep concerns which people who relate so closely with natural resources have about the intrinsic relationships and bonds between humans and the environment.

Hope and trust even in disaster

The fourth occasion of learning and unlearning came in Indonesia.

I was visiting the province of Aceh in 2007 after it had been devastated by the 2004 tsunami. I had heard about the separatist Free Aceh Movement and their bitter fight for the control of their natural resources and the new law for total provincial autonomy which they received almost like a 'gift of the tsunami'. I was intending to take on an assignment with the UN and this visit was part of an exploratory inception mission.

We were in the village of Patek in the district of Aceh Jaya which suffered the biggest loss of human life. It was just about 100 km from the epicentre of the gigantic (9 points on the Richter scale) underwater earthquake which triggered the massive tsunami. Practically the whole of Patek was wiped out. The greatest loss was among women, children and the elderly who were going about their daily morning chores on that fateful Sunday morning of 26 December 2004. Only the few fishers who were at sea on their boats survived. Pak Shaiffudin was one of them.

We met at the wayside family-run coffee shop – called the *kadai kopi* (coffee shop). Interestingly, I later came to know, this was the first establishment that was revived after a disaster in most villages. The *kadai kopi* is the main social institution, albeit a male-dominated one, where the newspaper is read aloud; where politics is heatedly discussed; where notes are compared on matters of the fishery; where economic and social deals are made.

I introduced myself to the fishers and followed the customary practice of shaking hands with every person seated in the coffee shop. The atmosphere was laden with cigarette smoke – a major weakness of Acehnese men. Being from India gave me preferential access to the group for two reasons: First, coastal Acehnese affirm they are all originally from India (Gujarat, West Bengal, Kerala and Tamil Nadu) and second, their inordinate love of Bollywood.

We entered into discussions on a wide range of topics – the latest of Bollywood movies and Shahrukh Khan's future; the fishery recovery and the flood of aid for rehabilitation; the 30-year war of the Free Aceh Movement with the brutal Indonesian army; and finally the details of that fateful day of the tsunami when about 180,000 people perished in a matter of 30 minutes.

I was surprised that the discussion on the 30-year war of Aceh was rife with such strong bitterness against the Indonesian armed forces and the majority ethnic communities' attitude to the Acehnese people. They were shocked at the brutality and injustice of fellow Indonesians towards them, leading to a loss of about 200,000 lives over the three decades.

And to my surprise, when we discussed the 30 minutes of the tsunami there was no special sense of grief or remorse. Some did seem very grieved while talking about lost loved ones. It seemed to me that for this group of survivors the universality of the losses became a source of mutual consolation.

I asked if in the wake of the terrible tragedy wrought on them by Nature the huge international aid effort had come as solace for their loss. A few of the fishers gave brief answers.

At the end Pak Shaiffudin spoke up. He said he had not yet decided what to do with his new home which was built for him by an aid agency. He had started to go back fishing over a year before. But the catches were not very good. The sea had changed radically. It was not giving up its wealth as it had done before at the depths and over the terrains that they were familiar with. There was a sense of magnanimous calmness in his manner of speaking. He concluded with this momentous statement, "The tsunami was not God's punishment but God's training".

I was so totally humbled by his indomitable faith and hope. My admiration and respect for him grew when I was later told that he was the one who suffered the greatest personal loss in the village – his wife, two children, parents, home and pets were devoured by the giant wave. This gave me the prime motivation to accept a four-year assignment in Aceh, which in UN terminology was a high-risk, insecure, hardship station. And Pak Shaiffudin became my great friend.

Here we are, scientists and climate activists meeting in big international conferences discussing the effects of climate change and the implications of the 1 metre rise in sea level with fear of the future, and Pak Shaiffudin talks about a 15-metre wave which devoured all that he could call his own as God's training – without remorse or bitterness! For fishers like Pak Shaiffudin, living in Aceh where earthquakes and tsunami threats are a regular phenomenon, the casualties of extreme events are intrinsically integrated into the expectations of a life made from living by and from the sea. Alluding to the armed conflict in Aceh, what they found more unpredictable and fearful were the unreasonable attitudes of their own fellow humans towards them!

Reflections

I am sure you perceive from the these encounters how very differently the fishers think about natural resources, the aquatic environment and their intrinsic relationship with them.

For sociologists, social scientists and reluctant academics like me, the challenge is to discover, in this alternate discourse of ecosystem people, new meanings and alternative ways of knowing and relating to Nature, which often come from people at the margins.

I am sure there are a variety of meanings which one can attribute to each of my narratives. Therefore, I make no claim that the reflections I make about them are unique or the only ones possible.

Consider the following:

> All the sayings have a certain directness and 'first person' approach. Undoubtedly the statements and questions are both excitingly poetic and deeply philosophical. The perspectives implied in these wise sayings arise from an intense lived experience, embodied meaning, material exchange and subjectivity. They are almost bursting at the seams pronouncing that however we look at it, we humans are always an integral part and parcel of Nature – whether Nature gives or takes away.

The narratives resoundingly echo the collective voice of individuals and communities who are generally 'invisible' to the neoliberal ideology because they may not be active consumers but just silent producers. It highlights their refusal to embrace discourses, goals and worldviews that are not innately their own. It points to their refusal to be cowed down by promises that everything will be okay if only they conform to the invisible hand of the market or the visible fist of the state.

I have four specific reflections about my narratives:

1 Pathrose Gomez's rhetorical question taught me that the degree of our connectedness with natural resources and the environment alters the way we view the *risks* of our relationship to them.
2 Joyachen Antony's quiet determination showed me that the degree to which our future is intertwined with resources and the environment determines the forthrightness of *resistance* we are willing to exert to protect it.
3 Kim Soeun's gentle rebuke highlighted how we tend to assess the gains from nature largely on the value of materials and services obtained from it. But it is the *rights and freedom of access* to natural resources without fear which are the *real* benefit.
4 Pak Shaifuddin's tranquil courage reminded me that Nature always surprises us. The way we accept this *innate uncertainty* is a function of our cognitive, affective and behavioural relationship with it.

In my limited understanding, I find these reflections and learnings from an ecosystem community counter-intuitive to the logic and rationale of the neoliberal agenda that is increasingly ordering our lives.

Their closeness to Nature and their intimate dependence on natural resources create the abiding faith in Nature's bounty. This faith provides ecosystem communities the courage to resist with fortitude the ill-conceived actions by biosphere communities that threaten to disrupt the flow of resources and the equilibrium of the environment. Much of the latter actions happen due to use of inappropriate technology and extracting excessive throughputs. Only such unambiguous assertion by ecosystem communities will ensure that they can obtain the unhampered freedom of access to resources and the environment which form the basis of their ability to lead a wholesome and sustainable livelihood. This is a state of affairs which many communities are striving not to lose, and few have even regained after bitter struggles. However, those who still have mastery over their resources and environment are acutely aware of the innate and totally unpredictable vicissitudes of Nature over which they have no control.

On the other side, we as 'biosphere' people are encouraged to detach ourselves and look at Nature 'dispassionately', to view it as 'separate from ourselves' and 'as the realm for natural capital resources and services' which enhance our luxury and lifestyles. We are assured that as individuals, we can reduce the risks and uncertainty associated with Nature if we can make a proper valuation, have the right technology and appropriate time perspective and credible information. Once this state is achieved we can take total control of Nature and use its resources to fashion a future of our own reckoning.

Is not the concrete evidence from around the world, of the way the environment is 'responding' to such a mainstream approach that is practised by the forces which control our lives today, enough to prove that we may indeed be dangerously wrong?

Perhaps then there is prescience in the voices of the many ecosystem people around the world who are increasingly speaking in one collective voice and challenging us 'biosphere' people to change our relationship with Nature and between ourselves.

I personally experienced hope in their voices.

But in the Vaclav Havelian sense where "Hope is an orientation of the spirit, and orientation of the heart; it transcends the world that is immediately experienced and is anchored somewhere beyond its horizons . . . it is definitely not the same thing as optimism. It is not the conviction that something will turn out well, but the certainty that something makes sense, regardless of how it turns out" (Havel 1990).

Notes

1 The term 'ecosystem people' – those who depend primarily on Nature for life and livelihood – and later in the text, the term 'biosphere people' – those who depend on all the resources of the planet – were first introduced by Dasmann (1976).
2 These are artisanal fishing boats made by tying logs of wood together.
3 For an elaboration of this 'outlier thesis' see Kurien (2000).
4 For details of this radical action see Kurien et al. (2006).

References

Dasmann, R.F. 1976. "Future Primitive: Eco-System People Versus Biosphere People," *Coevolution Quarterly* (Fall), 26–31.

Havel, Vaclav. 1990. *Disturbing the Peace: A Conversation with Karel Huizdala*, Translation by Paul Wilson, New York, Alfred A. Knopf.

Kurien, J. 2000. "The Kerala Model: Its Central Tendency and the 'Outlier'," in Parayil, G. (ed.): *Kerala: The Development Experience*, pp. 178–197, London, Zed Press.

Kurien, J., So, N. and Mao, S.O. 2006. *Cambodia's Aquarian Reforms: The Emerging Challenges for Policy and Research*, Phnom Penh, Cambodia, Inland Fisheries Research and Development Institute Publications.

INDEX

9780367553180